CHEETAH
Chasing and Capturing
Your Dreams with You

This book belongs to

_____

Who wants to become

A/An_____

By (enter date) _____

Student's signature_____

Date signed _____

Parent's signature_____

Date signed_____

CHEETAH
Connect to Higher Education, Electronic Tools, Aplication and Help

10 9 8 7 6 5 4 3 2

First published 2025 with an exclusive licence licence from the author to CHEETAH® Purrrrrrr Publishing ('CHEETAH®'), an imprint of CHEETAH® Toys & More, LLC.

ISBN-13: 978-1-964243-68-9

ISBN-10: 1-964243-68-9

Permission request(s) should be submitted to the publisher in writing at one of the addresses below:

Contact information:

CHEETAH® Toys More, LLC.
1382 Albany Ave, 2nd Floor,
Hartford CT 06112

Lot 138, Bryan's Bay
Port Antonio P.O.
Portland, Jamaica

info@mycheetahacademy.com
paulettetrowers@yahoo.com
876-909-6311 (WHATSAPP ONLY)
Author: Rhoen Kerr
Editor: Fiona Porter-Lawson
Reviewer: Kristina Jazz
Book and interior design: Feri Setiawan
Publisher: CHEETAH®

Contents

CHEETAH™
Connect to Higher Education, Electronic Tools, Aplication and Help

Dear CHEETAH family,

Preparing the Grade 4 Jamaican Mathematician is not a typical textbook. It is a combination of a textbook and a workbook, seamlessly teaching mathematical concepts while also providing ample opportunities for students to practice. This innovative approach ensures that students not only learn the necessary skills but also apply them in various exercises, reinforcing their understanding through hands-on activities and real-world problem solving. Additionally, there are several special features.

Special Features:

1. **100% Compliance with the new National Standards Curriculum (NSC):**

   o Ensures that all curriculum objectives are addressed in alignment with Jamaica's education standards.

2. **CAPE Approach (Key Concepts, Application, Practice, and Evaluation):**

   o **Key Concepts:** Clearly defined mathematical concepts are introduced in a structured format.

   o **Application:** Students apply these concepts through real-world examples and exercises.

   o **Practice (P in CAPE):** Various practice problems reinforce learning and mastery.

   o **Evaluation:** Students assess their understanding through quizzes, review questions, and reflection sections.

3. **Focus Questions:**

   o Each chapter starts with **focus questions** that set the learning objectives.

   o Encourages critical thinking and inquiry-based learning.

> **Let us practice 1.2**
>
> Write these numbers in their expanded form.
>
> a. 87 =
>
> b. 296 =
>
> c. 704 =
>
> d. 2,978 =
>
> e. 9,043 =

> **Focus question**
>
> How do I know the value of a number?

CHEETAH™
Connect to **H**igher **E**ducation, **E**lectronic **T**ools, **A**plication and **H**elp

4. **Prior Learning as a Diagnostic Tool:**

   o Includes **practice questions that act as a diagnostic gauge** for teachers.

   o Reinforces and reminds students of foundational information.

   o Supports students who may have missed or struggled with prior math concepts.

> **Prior learning:**
> ✓ Identify equivalent sets.
> ✓ Tell the worth of a set of notes and coin.

5. **Key Vocabulary:**

   o Each chapter includes a **glossary of important terms**.

   o Students are encouraged to **check their understanding** of these terms before moving forward.

> **Key vocabulary**
> Check the words you understand:
> ☐ census
> ☐ digit
> ☐ expanded form
> ☐ face value
> ☐ largest
> ☐ least
> ☐ place value
> ☐ standard form

6. **Answers & Explanations:**

   o Detailed solutions to exercises help students **self-check and correct errors**.

   o Step-by-step explanations are provided for complex problems.

7. **Curriculum Drivers & Engagement:**

   o Content is aligned with the **Jamaican Grade 4 Curriculum**.

   o The use of **Barky Bark, the Jamaican Math Detective**, makes learning more enjoyable. LaChase the CHEETAH leader is also a guest in this book!

   o **Colorful images and engaging activities** enhance comprehension and retention.

8. **Real-World Problem Solving:**

   o Questions are based on **practical scenarios**, such as:

      ▪ Calculating costs in Jamaican dollars.

- Measuring distances.

- Using multiplication and division in shopping scenarios.

- Applying geometry in everyday settings.

9. **Evaluation & Self-Reflection:**

   o Each unit ends with a **learning outcomes checklist** where students assess their progress.

   o Reflection activities encourage **metacognitive skills** (thinking about their thinking).

Learning outcome	No!	Working on it	Yes!
Did you get all the answers?	☹	😐	☺
Did you get most questions right?	☹	😐	☺
Did you retry the question(s) you got wrong?	☹	😐	☺
Were you able to correct your wrong answers?	☹	😐	☺
If not, did you seek help from others and/or review the chapter?	☹	😐	☺

10. **ICT Integration – Available as Interactive E-Books and Teaching Videos:**

- Supports **digital learning** with interactive e-books.

- Includes **teaching videos** to enhance understanding and engagement.

This **interactive, student-centered workbook** provides **a structured, engaging, and culturally relevant** approach to learning mathematics. By integrating **Barky Bark**, **real-life applications**, **diagnostic tools for teachers**, and **ICT resources**, it ensures students **develop mathematical fluency and problem-solving skills** in a fun and effective way.

I am Bark, Barky Bark, that is, the Jamaican math detective. I will educate, entertain and inspire you the CHEETAH™ way. Let's solve some math problems.   Are you ready?
Let's go. Let's prep for life!

CHEETAH™
Connect to **H**igher Education, **E**lectronic **T**ools, **A**plication and **H**elp

TERM 1

A. 235 + 123

B.   1/4 , 3/4 , 2/4 , 4/4

How do I know the value of a number?

Before we begin, let's see what you know.

Prior learning:

✓ Read and write 4-digit numbers.
✓ Partition and combine groups of objects.
✓ Identify fractional numbers (halves to tenths).
✓ Compute with whole numbers (up to 3 digits).
✓ Round whole numbers to the nearest thousand.

Key vocabulary

Check the words you understand:

☐ census
☐ digit
☐ expanded form
☐ face value
☐ largest
☐ least
☐ place value
☐ standard form

1. Read aloud then write the number 7,563 in words.

_____

2. Partition (break up) and combine groups of numbers to tell if these number partitions are true or false. Tick the correct box.

Number	Partition	True	False
20	10 + 5 + 2		
45	5 + 20 + 10 + 3 + 7		
48	12 + 12 + 12 + 6 + 4 + 2		

3. Identify the fraction shown by the shaded region (halves to tenths). Write the fractions name and value on the line given.

a. _____   b. _____   c. _____   d. _____   e. _____

f. _____   g. _____   h. _____

3. Complete the following tasks. Add or subtract these numbers.

a. 235 + 123   b. 687 + 254   c. 236 - 123   d. 568 - 379   e. 600 - 275

$$\begin{array}{r} 235 \\ + 123 \\ \hline \end{array}$$

$$\begin{array}{r} 236 \\ - 123 \\ \hline \end{array}$$

$$\begin{array}{r} 568 \\ - 379 \\ \hline \end{array}$$

$$\begin{array}{r} 600 \\ - 275 \\ \hline \end{array}$$

4. Multiply or divide the following.

a. $\begin{array}{r} 23 \\ \times 4 \\ \hline \end{array}$ b. $\begin{array}{r} 47 \\ \times 6 \\ \hline \end{array}$ c. $\begin{array}{r} 58 \\ \times 5 \\ \hline \end{array}$ d. $484 \div 3$ e. $603 \div 3$ f. $515 \div 5$

If you know these, you should be able to learn what comes next.

5. Round off these whole numbers to the nearest thousand.

a. 1234 =   b. 468 =
c. 1573 =   d. 2670 =

🎯 I.I Distinguish between the face value, place value and true value of a digit in a number.

What is a number?

A number is an amount or quantity of something. It answers the question how many or how much. A number is used in mathematics and science to count, measure and label.

Where do we get these numbers from?

We use the Hindu-Arabic numeral (number) system which uses a set of 10 symbols (**digits**). The 10 digits are: 0, 1, 2, 3, 4, 5, 6, 7, 8, 9. These are the digits in the decimal number system.

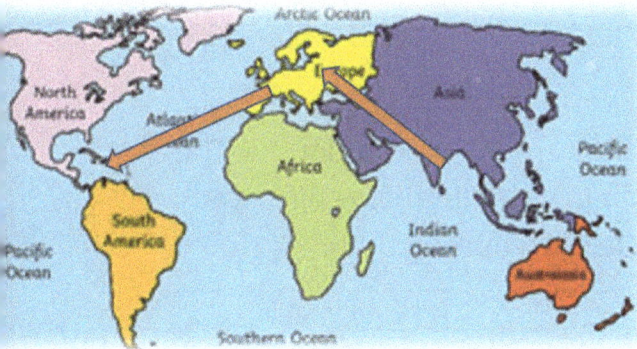

These digits were first used in India in the 6th or 7th century, then brought to the Americas through the writings of Middle Eastern mathematicians.

What is the face value of a digit in a number?

The **face value** of a digit is how the number 'looks'. Yes! Right! For example, this number is a five (5). When you look at it, you note its shape and can say it is a five. The digit 5 has its own face and is written differently from other digits which also have their own faces: 0, 1, 2, 3, 4, 6, 7, 8, 9.

What is the place value of a digit in a number?

When we see a number that is greater than 9, it is set out in a row which we read from left to right. This is called the **standard form** of the number. Example: 7,563. Each digit in the number 7563 has its own value based on where the digit is in the number.

CHEETAH
Connect to Higher Education, Electronic Tools, Aplication and Help

Place value chart

The **place value** of a digit is where the digit is placed or positioned in a number. Let's look at how we read the place value of each digit in the number 7563.

Millions	Hundred thousands	Ten thousands	Thousands	Hundreds	Tens	Ones
			7	5	6	3

X 10   X 10   X 10   X 10   X 10   X 10

largest ← → smallest

increasing by 10

Here 3 is under the ones, 6 is under the tens, 5 is under the hundreds, and 7 is under thousands. The value of a digit changes when the place value of the digit changes.

What is the true value of each digit in the number 7563?

The true value of a digit depends on the digit's position in the place value chart. The true value of a digit is its face value multiplied by its place value. Here 3 is 3 x 1, 6 is 6 x 10, 5 is 5 x 100 and 7 is 7 x 1000. The value of a digit can change when the place value of a digit changes.

The true value of the digit 3 is 3, 6 is 60, 5 is 500 and 7 is 7000. So, 3 + 60 + 500 + 7000 = 7563

Example: Look at these three numbers below. They are different amounts. The face value of five is the same. The place value of 5 in each number is different. This causes the true value of five to be different in each number.

		True value of 5
1. (5)9 3	5 is at the hundred place or location →	5 x 100 = 500
2. 9(5)3	5 is at the tens place or location →	5 x 10 = 50
3. 9 3(5)	5 is at the ones place or location →	5 x 1 = 5
Note: The true value of a digit is the face value multiplied by the place value.		

Let us practice l.l

Find the face value, place value and true value of the digits in **bold**.

26	675	1,3**9**8
face value _____	face value _____	face value _____
place value _____	place value _____	place value _____
true value _____	true value _____	true value _____
38,456	**2**03,685	9,7**5**4,219
face value _____	face value _____	face value _____
place value_____	place value _____	place value _____
true value _____	true value _____	true value _____

🎯 **1.2 & 1.3** Identify the value of whole numbers with up to seven digits. Read and write whole numbers with up to seven digits.

What is the value of each digit in a whole number?

The value of the digits in a whole number can be seen in the **expanded form**.

Number	Expanded form
26	2 x 10 + 6 x1
345	3 x 100 + 4 x 10 + 5 x 1
5,306	5 x 1000 + 3 x 100 + 0 x10 + 6 x 1
8,675,234	8 x 1,000,000 + 6 x 100,000 + 7 x 10, 000 + 5 x 1000 + 2 x 100 + 3 x 10 + 4 x 1

Let us practice 1.2

Write these numbers in their expanded form.

a. 87 = _____

b. 296 = _____

c. 704 = _____

d. 2,978 = _____

e. 9,043 = _____

f. 370,864 = _____

g. 697,325 = _____

h. 7,598,620 = _____

i. 8,659,423 = _____

j. 4,097,351 = _____

How do we read and write a number?

The place value chart is used to read numbers with more than one digit such as 94, 594, 1594, 81594, 681594 and 3681594. These numbers can be used to count the number of items or people in a government **census**, the money we have or even the number of houses on a street.

Let us look at the number 3681594 on a place value chart.

Millions	Hundred thousands	Ten thousands	Thousands	Hundreds	Tens	Ones
3	6	8	1	5	9	4
three million	six hundred and eighty-one thousand			five hundred	ninety-four	

When we write a number, we write the digits **largest** to smallest in value. To help in reading the number, a comma (,) may be placed in front of every 3 digits counted from the right of the number.

So, the number in words is three million, six hundred and eighty-one thousand, five hundred and ninety-four. In standard form, this number is written as 3,681,594.

Let us practice 1.3

1. Change these numbers from standard form into worded form.

Standard form	Worded form
a. 2,469	
b. 3,876	
c. 28,964	
d. 39,453	
e. 826,351	
f. 753,826	
g. 1,011,011	
h. 1,204,507	
i. 2,003,014	
j. 9,876,543	

2. Change these numbers from words to their expanded form, and the expanded form to words.

Worded form	Expanded form
a. four hundred and seventy-six	
b. eighty thousand, four hundred and twenty-five	
c. twenty-six thousand eight hundred and eight	
d. two hundred and twenty-eight thousand, six hundred and seventy	
e. six million, seven hundred and five thousand and two	
f.	$2 \times 10 + 4 \times 1$
g.	$8 \times 100 + 4 \times 10 + 6 \times 1$
h.	$7 \times 1,000 + 3 \times 100 + 6 \times 10 + 9 \times 1$
i.	$8 \times 10,000 + 5 \times 1,000 + 8 \times 100 + 6 \times 10 + 5 \times 1$
j.	$6 \times 100,000 + 0 \times 10,000 + 2 \times 1,000 + 0 \times 100 + 6 \times 10 + 7 \times 1$
k.	$8 \times 1,000,000 + 7 \times 100,000 + 4 \times 10,000 + 2 \times 1,000 + 8 \times 100 + 6 \times 10 + 5 \times 1$

Remember to ask for help if you need it.

CHEETAH
Connect to Higher Education, Electronic Tools, Aplication and Help

3. Change these numbers from words to their standard form.

Worded form	Standard form
a. four hundred and seventy-six	
b. eighty thousand, four hundred and twenty-five	
c. twenty-six thousand eight hundred and eight	
d. two hundred and twenty-eight thousand, six hundred and seventy	
e. six million, seven hundred and five thousand and two	

4. Use the internet to investigate which of these numbers will be bigger. Place a tick (✓) in the correct column.

Column A	Column B
cost of buying a car	cost of buying a bus
distance from Kingston to Ocho Rios	distance from your home to your school
the population of Jamaica	population of Kingston
distance from Earth to the moon	distance from Earth to the Sun
the amount of money needed to buy groceries for one week at home	the amount of money needed to buy a school's canteen supplies for one week

Evaluation: Let us see how you did.

Learning outcome	No!	Working on it	Yes!
Did you get all the answers?	☹	😐	🙂
Did you get most questions right?	☹	😐	🙂
Did you retry the question(s) you got wrong?	☹	😐	🙂
Were you able to correct your wrong answers?	☹	😐	🙂
If not, did you seek help from others and/or review the chapter?	☹	😐	🙂

Colour the face that shows how you are doing.

What do I need to know about sets?

Prior learning:
- ✓ Identify equivalent sets.
- ✓ Tell the worth of a set of notes and coin.

Before we begin, let's see what you know.

Key vocabulary

Check the words you understand:

- ☐ attributes
- ☐ characteristics
- ☐ differences
- ☐ groups
- ☐ members
- ☐ sets
- ☐ similarities
- ☐ sorting

1. Complete the following table on ideas about sets.

	Questions	Answer
i.	What does { } mean?	
ii.	If A = { 2, 4, 6}, what are the elements in the set?	
iii	Describe set A in words.	
iv	What is the name given to sets with the same number of members? Example A = { 1, 2, 3} and B = { 4, 5, 6}	

2. Tom opened his saving pan and showed his sister his coins.

$$S = \{ 1\$, 1\$, 10\$, 10\$, 10\$, 20\$, 20\$ \}$$

Count and tell how much Tom saved: _____

3. Mary went to the market. When she opened her purse, she saw the set of dollar notes below.

$$\text{Set A} = \{ 100\$, 50\$, 50\$, 100\$, 100\$, 50\$ \}$$

Add the dollar notes and tell how much money she had in her purse: _____

If you know these, you should be able to learn what follows.

🎯 2.1 Define the concept of a set.   2.2 Describe a set.
2.3 Name any set using braces.   2.4 Name and list members of any given set.

2.1 Define the concept of a set.

A **set** is a collection or **group** of objects, things, items, ideas or symbols that have something or things in common (**similarities**). It may be a group of numbers, shapes, people, letters of the alphabet, days of a week, types of vehicles, and so on. Sets are used to **sort** things into groups.

2.2 Describe a set and name any set using braces.

Every item in the set is called an element or a **member** of the set. When we describe a set, we state the group to which the members belong. What makes up a set shows its **characteristics** or **attributes** and can be used to guess the missing members in a set. Example: Set A is the set of counting numbers from 1 to 5.

2.3 Name any set using braces

All sets are written in a similar way. Curly brackets (braces) are used to keep the members of a set together. Example: Set A = {1,2,3,4,5}. Sets are named using a capital letter before the brackets. Members of the set are listed inside the brackets. They are separated by a comma (,).

2.4 Name and list members of any given set; Ways to describe a set of elements.

A set can be written in words or using set notation. Set notation lists the members of any given set and can show the number of members in the set. Where there are three dots (...) this means the number of elements in the set cannot be counted. We call this infinite. For example, the set of whole numbers, W = {0,1,2,3,4...).

Here are the **differences** in how we show the information for a set.

In words	Using set notation
Set M = the first 10 whole numbers	Set M = { 0, 1, 1, 3, 4, 5, 6, 7, 8, 9}
The number of elements in a set	
n(M) = 10	

Let us practice 2.1–2.4

1. Here are some sets. Describe the sets in words and count the number of elements in the sets.

	Members	In words	Number of elements in the set
a.	Set P = [m, n, o, p, q, r, s, t}		n(P) =
b.	Set K = {knives, forks, spoons, cups, saucer, plates}		n (K) =
c.	Set N = {1, 2, 3, 4, 5, 6, 7, 8, 9}		n(N) =

2. Describe the following sets in words.

a.	Set V = {a, e, i, o, u}	
b.	Set M = {shirts, pants, jackets, dresses, blouses}	
c.	Set Y = {tables, chairs, sofas}	
d.	Set T = {plywood, hammer, screwdriver, drill, saw}	
e.	Set C = {pig, cow, goat, chicken}	

3. Which option tells the meaning of a set in mathematics?
   a. A set is the name of things.
   b. A set is what comes before a race starts.
   c. A set is a group of things in a bracket.
   d. A set is a group of things.

4. Select the set which is correctly written.
   a. Set R = {l, 2, 3
   b. Set R = {2. 3. 6}
   c. Set R = 5, 6, 7}
   d. Set R = {2, 5, 9]

5. What does this symbol { } tell about a set?
   a. There are only a few elements in the set.
   b. There are no elements in the set.
   c. The set is broken.
   d. The members of the set are invisible.

6. In set B = {l, 2, 3, 4 …}, what is the meaning of the 3 dots?
   a. The set is full.
   b. The set has 4 members.
   c. The set is empty.
   d. The members of the set cannot be counted.

7. State the n(F) = _____, if set F is the number of different types of fruits in the picture.

8. A student was asked to list items that go into the set of objects used in the classroom. The student named the set C. Write the element that is out of place and state why you think it is out of place by writing a set description in words.

Set C = {pencil, pen, paper, book, glue, electric iron}

Out of place:_____

Explain: _____

9. A child at a daycare sorts the shapes shown below into different sets.

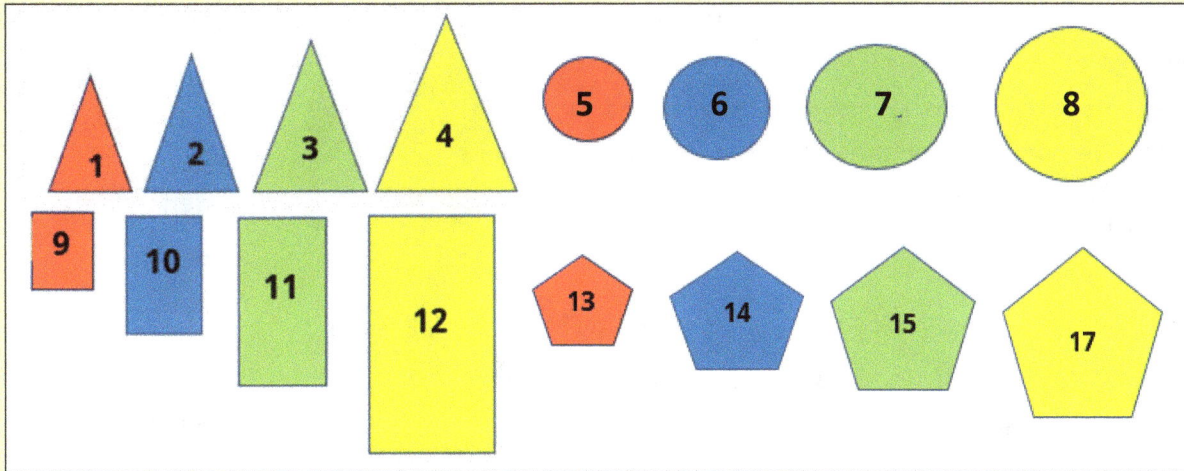

Create a set (S) with these members using the numbers on each shape.			
a. set of all triangles		g. set of all yellow shapes	
b. set of all circles		h. set of all green shapes	
c. set of all rectangles		i. set of all the smallest shapes	
d. set of all pentagons		j. set of all the largest shapes	
e. set of all red shapes		k. set of all shapes	
f. set of all blue shapes		l. set of all pink shapes	

Keep practising and you will succeed.

Evaluation: Let us see how you did.

Learning outcome	No!	Working on it	Yes!
Did you get all the answers?	☹	😐	☺
Did you get most questions right?	☹	😐	☺
Did you retry the question(s) you got wrong?	☹	😐	☺
Were you able to correct your wrong answers?	☹	😐	☺
If not, did you seek help from others and/or review the chapter?	☹	😐	☺

Colour the face that shows how you are doing.

CHEETAH™
Connect to **H**igher Education, **E**lectronic **T**ools, **A**plication and **H**elp

How do I apply fraction ideas to real life situations?

Before we begin, let's see what you know.

Prior learning:

✓ Name parts of fractions, i.e., halves through tenths.
✓ Identify the numerator or denominator in a fraction.
✓ Place fractions with same denominator/numerator in serial order.
✓ Write fraction families.

Key vocabulary

Check the words you understand:

☐ denominator
☐ equivalent
☐ estimate
☐ improper fractions
☐ numerator
☐ parallelogram
☐ proper fractions
☐ unit fraction

A fraction is a part of a whole.

A fraction is less than a whole.

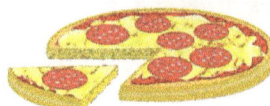

A fraction is written in two parts. The top number is called the **numerator**, and the bottom number is called the **denominator**. Three-quarters (3/4) is an example of a fraction.

The numerator tells the number of parts you have out of the whole. The denominator tells the number of equal parts the whole is divided into. So, ¾ of a cake means we have 3 parts out of the 4 parts the cake is divided into.

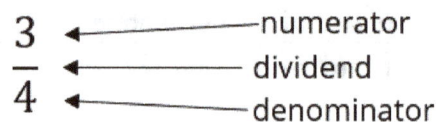

$$\frac{3}{4}$$

← numerator
← dividend
← denominator

1. Write the name of the fractions in the table below.

a. _____	b. _____	c. _____	d. _____	e. _____	f. _____

2. Fraction families are fractions with the same denominator.

Fractions families	Place fraction families in ascending order
a. $\frac{3}{3}, \frac{1}{3}, \frac{2}{3}$	
b. $\frac{1}{4}, \frac{3}{4}, \frac{2}{4}, \frac{4}{4}$	
c. $\frac{4}{5}, \frac{3}{5}, \frac{2}{5}, \frac{1}{5}, \frac{5}{5}$	
d. $\frac{5}{7}, \frac{3}{7}, \frac{2}{7}, \frac{6}{7}, \frac{1}{7}, \frac{4}{7}$	

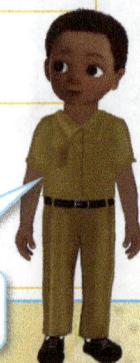

If you know these, you should be able to learn what comes next.

3.1 Distinguish among whole numbers, proper fractions, improper fractions and mixed numbers.
3.2 Recognize like fractions.
3.3 Write fractions in their simplest forms using the understanding of equivalence.
3.4 Order fractions with different denominators.
3.5 Benchmark fractions using 0, ½ and 1.
3.6 Subtract a proper fraction or a mixed number from a whole number.

3.1 Distinguish among whole numbers, proper fractions, improper fractions and mixed numbers.

Whole numbers

Whole numbers (W) are the set of counting numbers and 0.
Example: W = {0, 1, 2, 3...}

whole $\dfrac{4}{4}$

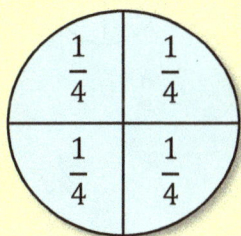

$$\frac{4\ parts}{4\ parts} = 1\ whole$$

Proper fractions

When the numerator is smaller than the denominator, the fraction is called a **proper fraction**. Examples of proper fractions are $\dfrac{1}{2}, \dfrac{3}{4}, \dfrac{5}{6}, \dfrac{4}{9}, \dfrac{7}{12}, \dfrac{13}{15}$ and so on as long as the numerator is smaller than the denominator.

fraction $\dfrac{3}{4}$

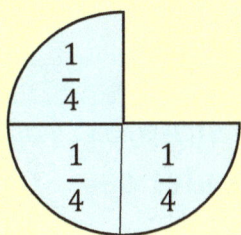

$$\frac{3\ parts}{4\ parts}$$

Improper fraction

When the numerator is larger than the denominator, the fraction is said to be an **improper fraction**.

E.g.: $\dfrac{7}{3}, \dfrac{9}{2}, \dfrac{12}{7}, \dfrac{5}{2}, \dfrac{7}{2}, \dfrac{8}{3}, \dfrac{6}{5}$

Mixed number

A mixed fraction is a number that has a whole number and a proper fraction.

Example: colouring the mixed fraction $1\dfrac{3}{4}$

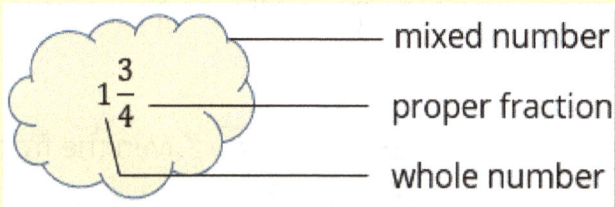

$1\dfrac{3}{4}$ — mixed number
— proper fraction
— whole number

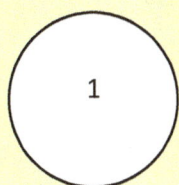
1

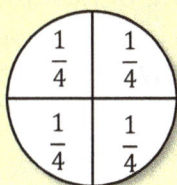

gives

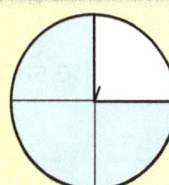

CHEETAH™
Connect to Higher Education, Electronic Tools, Aplication and Help

Let us practice 3.1

Shade or colour the following mixed fractions.

a.

$1\frac{1}{2}$

b.

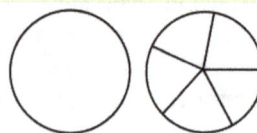

$1\frac{4}{5}$

c.

$1\frac{7}{8}$

d.

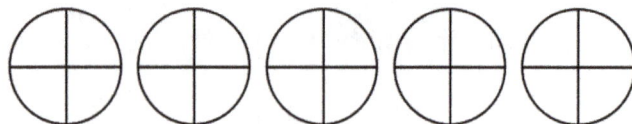

$2\frac{2}{3}$

e.

$4\frac{3}{4}$

3.2 Recognize like fractions.

Like fractions are fractions with equal denominators such as $\frac{1}{6}$ and $\frac{4}{6}$.

Example: Given $\frac{2}{4}, \frac{2}{3}, \frac{1}{4}, \frac{3}{7}, \frac{3}{4}, \frac{4}{6}, \frac{3}{5}$ list the like fractions.

Answer: $\frac{2}{4}, \frac{3}{4}, \frac{1}{4}$

3.3 Write fractions in their simplest forms using the understanding of equivalence.

Equivalent fractions are fractions that are equal in value but have different numerators and denominators.

> Note that two quarters make one half.

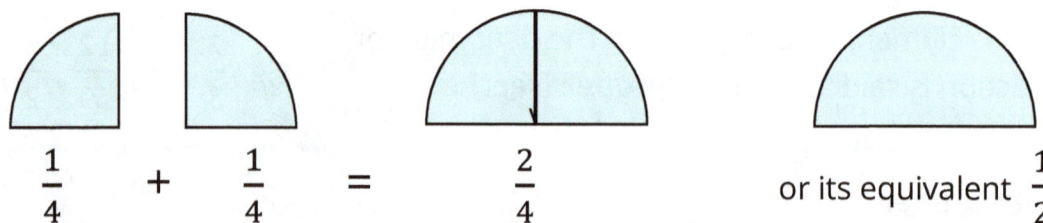

$\frac{1}{4}$ + $\frac{1}{4}$ = $\frac{2}{4}$ or its equivalent $\frac{1}{2}$

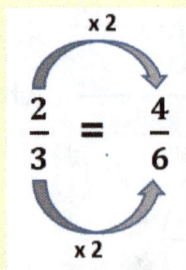

Given the fraction $\frac{2}{3}$, to get an equivalent fraction we can multiply the numerator and the denominator by the same number. Hence $\frac{2}{3}$ is equivalent to $\frac{4}{6}$.

3.1 Distinguish among whole numbers, proper fractions, improper fractions and mixed numbers.
3.2 Recognize like fractions.
3.3 Write fractions in their simplest forms using the understanding of equivalence.
3.4 Order fractions with different denominators.
3.5 Benchmark fractions using 0, ½ and 1.
3.6 Subtract a proper fraction or a mixed number from a whole number.

3.1 Distinguish among whole numbers, proper fractions, improper fractions and mixed numbers.

Whole numbers

Whole numbers (W) are the set of counting numbers and 0.

Example: $W = \{0, 1, 2, 3...\}$

whole $\dfrac{4}{4}$

$\dfrac{4\ parts}{4\ parts} = 1\ whole$

Proper fractions

When the numerator is smaller than the denominator, the fraction is called a **proper fraction**. Examples of proper fractions are $\dfrac{1}{2}, \dfrac{3}{4}, \dfrac{5}{6}, \dfrac{4}{9}, \dfrac{7}{12}, \dfrac{13}{15}$ and so on as long as the numerator is smaller than the denominator.

fraction $\dfrac{3}{4}$

$\dfrac{3\ parts}{4\ parts}$

Improper fraction

When the numerator is larger than the denominator, the fraction is said to be an **improper fraction**.

E.g.: $\dfrac{7}{3}, \dfrac{9}{2}, \dfrac{12}{7}, \dfrac{5}{2}, \dfrac{7}{2}, \dfrac{8}{3}, \dfrac{6}{5}$

Mixed number

A mixed fraction is a number that has a whole number and a proper fraction.

Example: colouring the mixed fraction $1\frac{3}{4}$

$1\frac{3}{4}$ — mixed number — proper fraction — whole number

1   gives

Let us practice 3.1

Shade or colour the following mixed fractions.

a.

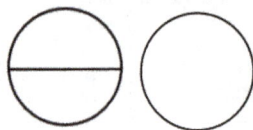

$1\frac{1}{2}$

b.

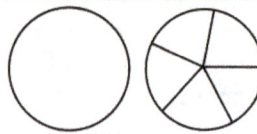

$1\frac{4}{5}$

c.

$1\frac{7}{8}$

d.

$2\frac{2}{3}$

e.

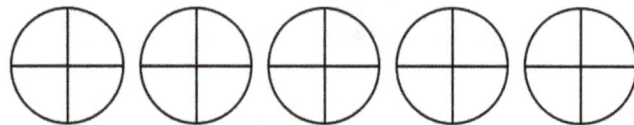

$4\frac{3}{4}$

3.2 Recognize like fractions.

Like fractions are fractions with equal denominators such as $\frac{1}{6}$ and $\frac{4}{6}$.

Example: Given $\frac{2}{4}, \frac{2}{3}, \frac{1}{4}, \frac{3}{7}, \frac{3}{4}, \frac{4}{6}, \frac{3}{5}$ list the like fractions.

Answer: $\frac{2}{4}, \frac{3}{4}, \frac{1}{4}$

3.3 Write fractions in their simplest forms using the understanding of equivalence.

Equivalent fractions are fractions that are equal in value but have different numerators and denominators.

> Note that two quarters make one half.

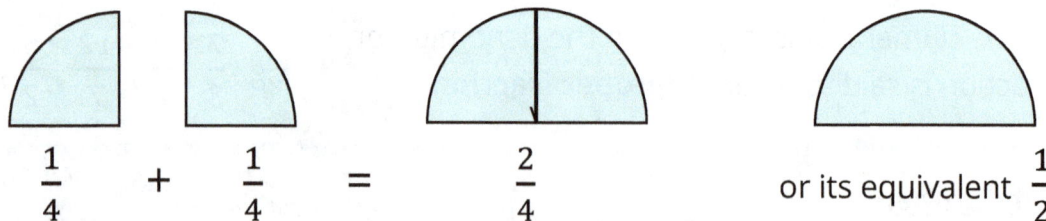

$\frac{1}{4}$ + $\frac{1}{4}$ = $\frac{2}{4}$ or its equivalent $\frac{1}{2}$

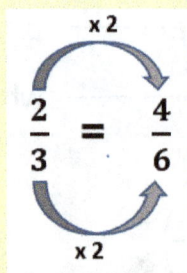

$\frac{2}{3} = \frac{4}{6}$

Given the fraction $\frac{2}{3}$, to get an equivalent fraction we can multiply the numerator and the denominator by the same number. Hence $\frac{2}{3}$ is equivalent to $\frac{4}{6}$.

CHEETAH
Connect to Higher Education, Electronic Tools, Aplication and Help

Given $\dfrac{15}{20}$, we can either multiply or divide the numerator and denominator to get equivalent fractions.

Hence $\dfrac{15}{20}$ is equivalent to both $\dfrac{3}{4}$ and $\dfrac{30}{40}$.

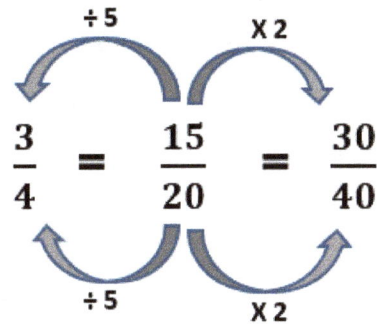

$$\dfrac{3}{4} = \dfrac{15}{20} = \dfrac{30}{40}$$

÷5   X2
÷5   X2

Let us practice 3.2 & 3.3

Make 2 equivalent fractions of each.

a. $\dfrac{1}{3} =$

b. $\dfrac{2}{5} =$

c. $= \dfrac{3}{6} =$

d. $= \dfrac{10}{20} =$

3.4 Order fractions with different denominators.

Given the task to arrange $\dfrac{1}{2}, \dfrac{1}{3}$ and $\dfrac{1}{4}$ in ascending order, we can follow the following steps.

Step 1: Find a common denominator:

Multiply all denominators: 2 x 3 x 4 = 24

Step 2: Convert all fractions to the common denominator:

Multiply each fraction by a number that makes their denominator 24.

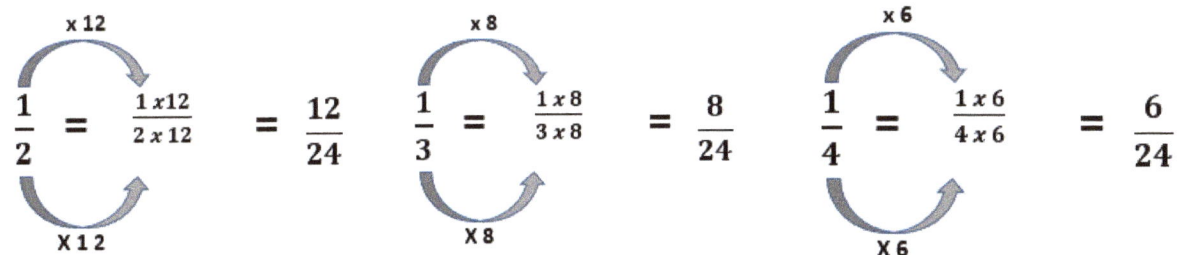

x 12
$$\dfrac{1}{2} = \dfrac{1 \times 12}{2 \times 12} = \dfrac{12}{24}$$
X 1 2

x 8
$$\dfrac{1}{3} = \dfrac{1 \times 8}{3 \times 8} = \dfrac{8}{24}$$
X 8

x 6
$$\dfrac{1}{4} = \dfrac{1 \times 6}{4 \times 6} = \dfrac{6}{24}$$
X 6

Hence, we have: $\dfrac{1}{2} = \dfrac{12}{24}, \quad \dfrac{1}{3} = \dfrac{8}{24} \quad and \quad \dfrac{1}{4} = \dfrac{6}{24}$

Ordering the like fractions gives $\dfrac{6}{24} < \dfrac{8}{24} < \dfrac{12}{24}$

Note that is $\dfrac{1}{4} < \dfrac{1}{3} < \dfrac{1}{2}$, so the order is $\dfrac{1}{4}, \dfrac{1}{3}$ then $\dfrac{1}{2}$.

When the denominator is common, the larger numerator makes the larger fraction.

We could also use 12 as a common denominator for

$$\dfrac{1}{4} < \dfrac{1}{3} < \dfrac{1}{2}$$

Lowest Common Multiple (LCM).

Let us practice 3.4

1. State which is larger by ordering the fractions $\frac{2}{3}, \frac{4}{5}, \frac{5}{6}$ giving them the same denominator.

2. Place these fractions in order from smallest to largest by converting to equivalent fractions.

   a. $\frac{2}{3}, \frac{4}{5}, \frac{1}{2}$

   b. $\frac{2}{3}, \frac{3}{4}, \frac{4}{5}$

   c. $\frac{3}{4}, \frac{1}{3}, \frac{1}{2}$

   d. $\frac{2}{3}, \frac{1}{2}, \frac{3}{5}$

3.5 Benchmark fractions using 0, ½ and 1.

To benchmark fractions means to show where the fraction falls between 0 and 1 on a number line.

Example 1:

Find the fraction marked by the position of A on the number line.

First, count the number of spaces between 0 and 1 to know the denominator for the fraction = 4.

Second, count the number of spaces between 0 to A to know the numerator for the fraction = 3.

Answer: Position of A is $\frac{3}{4}$

Example 2:

Find the fraction marked by the position of A on the number line.

First, count the number of spaces between 0 and 1 to know the denominator for the fraction = 8.

Second, count the number of spaces between 0 to P to know the numerator for the fraction = 6.

Answer: Position of P is $\frac{6}{8}$

Can you tell an equivalent fraction for $\frac{6}{8}$?

Let us practice 3.5

1. Read the number line at point A using fractions.

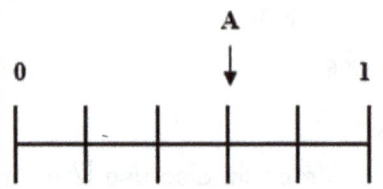

   A =

2. Read the number line at point G using fractions.

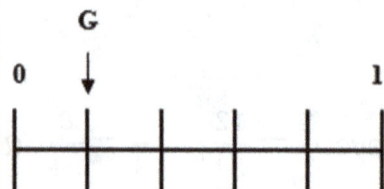

   G =

3. Read the number line at point T using fractions.

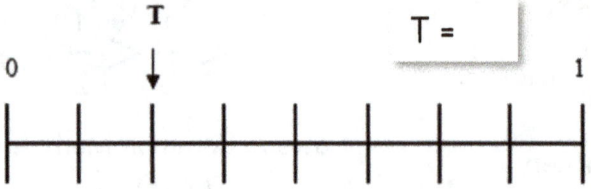

   T =

4. Read the number line at point N using fractions.

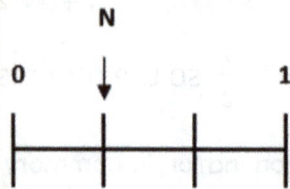

   N =

3.6 Subtract a proper fraction or a mixed number from a whole number.

Proper fraction from a whole

Example 1: Imagine that you have 2 apples and you gave half of one to a friend. How many apples do you have left?

So, we are finding 2 whole apples - ½ of an apple.

Note that 2 is equal to 4 halves: $2 = \frac{4}{2}$

$$2 - \frac{1}{2}$$
$$= \frac{4}{2} - \frac{1}{2}$$
$$= \frac{3}{2}$$
$$= 1\frac{1}{2}$$

Answer: 2 - ½ = 1½

Example 2: Suppose you have 3 candy bars and you wish to give your friend a quarter of one bar. How much candy bar would you have left?

Let's divide the 3 bars into quarters. Remember that each bar has 4 quarters because 4 quarters make one whole bar.

Also note that 3 = 2 + 1 or $3 = 2 + \frac{4}{4}$

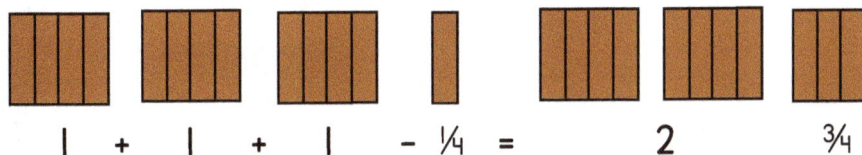

$$3 - \frac{1}{4}$$
$$= 2 + \frac{4}{4} - \frac{1}{4}$$
$$= 2 + \frac{4-1}{4}$$
$$= 2\frac{3}{4}$$

Answer: 3 – ¼ = 2 ¾

Let us practice 3.6.1

a. $4 - \frac{5}{7} =$ _____

b. $1 - \frac{2}{5} =$ _____

c. $3 - \frac{1}{3} =$ _____

d. $2 - \frac{5}{8} =$ _____

e. $1 - \frac{5}{9} =$ _____

f. $4 - \frac{7}{8} =$ _____

g. $3 - \frac{3}{5} =$ _____

h. $1 - \frac{3}{4} =$ _____

i. $2 - \frac{5}{6} =$ _____

j. $2 - \frac{3}{7} =$ _____

Mixed number from a whole number

Example: Four (4) sandwiches were made for a group of kindergarten students' lunches.

    a. How many of the sandwiches are left after 2 ½ are eaten.

    b. If 2 ½ sandwiches were eaten and each student got ½ of a sandwich, how many students got lunch?

a. Number of sandwiches left = 4 - 2 ½

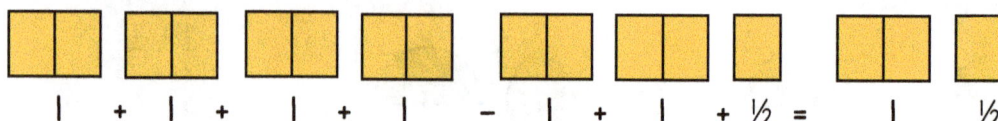

| + | + | + | − | + | + ½ = | ½ |

$$4 - 2½ = 1½$$

Hence, 1 ½ of the sandwiches is left.

$$4 - 2½$$
$$= \frac{8}{2} - \frac{5}{2}$$
$$= \frac{3}{2}$$
$$= 1\frac{1}{2}$$

b. If each whole sandwich was divided into 2 halves, then they have 5 half sandwiches.

| | | ½ |

Hence, 5 students got lunch.

Did you know that $2\frac{1}{2}$ can be converted to $\frac{5}{2}$?

Step 1: Multiply whole number by denominator.

Step 2: Add numerator to answer from step 1.

$$2 \begin{array}{c} +1 \\ \times \ 2 \end{array}$$

Let us practice 3.6.2

I.   Solve the following questions:

a.	$2 - 1½$	b.	$4 - 3\frac{4}{5}$
c.	$3 - 1\frac{2}{3}$	d.	$3 - 1\frac{3}{8}$
e.	$3 - 1\frac{2}{5}$	f.	$2 - 1\frac{3}{5}$
g.	$4 - 2\frac{3}{5}$	h.	$5 - 3\frac{2}{7}$
i.	$4 - 2\frac{5}{8}$	j.	$2 - 1\frac{7}{8}$

2. Solve the following fraction problems.

a. Nicholas had 4 chocolate bars and gave Shanique 2 1/3. How many did he have left?

_____

b. There are 5 ice pops. If each student got ½ of an ice pop, how many students got ice pops?

_____

c. Divide 4 buns equally among 8 people. How many does each get?

_____

d. Divide 3 ½ biscuits equally among 7 people. How many does each get?

_____

e. Share a cake among 3 people, so that no two person gets the same amount.

_____

Evaluation: Let us see how you did.

Learning outcome	No!	Working on it	Yes!
Did you get all the answers?	☹	😐	☺
Did you get most questions right?	☹	😐	☺
Did you retry the question(s) you got wrong?	☹	😐	☺
Were you able to correct your wrong answers?	☹	😐	☺
If not, did you seek help from others and/or review the chapter?	☹	😐	☺

Colour the face that shows how you are doing.

How can I estimate and verify my answers?

Before we begin, let's see what you know.

Prior learning:

✓ Mentally recall addition and subtraction of 2-digit numbers.
✓ Construct addition and subtraction problems.
✓ Use and write three-digit numbers in standard form.
✓ Round two-digit numbers to nearest ten.
✓ Use rounded numbers to estimate answers for addition and subtraction problems.

Key vocabulary

Check the words you understand:

☐ digit
☐ addition
☐ subtraction
☐ mentally
☐ estimate

Addition and subtraction of 2-digit numbers.

1. Work out the answer to these questions by doing them in your mind.

   a). 42 + 24   b). 58 + 21   c). 89 – 64   d). 51 - 24

Construct addition and subtraction problems.

2. a). There are 43 pencils in one bag and 25 pencils in another. How many pencils are in both bags altogether?

   _____

   b). Mary has $46 and Tony has $39. How much more do they need to buy chocolate which costs $90?

   _____

Writing three-digit numbers in standard form.

3. a). Change 684 into words and expanded form. _____

   b). There were eight hundred and seventy-four pages in the book. Write this number in standard form and expanded form.

   _____

4.

Round off these two-digit numbers to the nearest 10.		Use the rounding off method to do these questions mentally.	
a.	27	f.	45 + 72 =
b.	84	g.	28 + 64 =
c.	55	h.	86 - 24 =
d.	72	i.	97 - 48 =
e.	67	j.	26 -16 + 10 =

If you know these, you should be able to learn what comes next.

4.1 Make reasonable estimate when computing whole numbers.
4.2 State how the properties of commutativity and associativity apply to addition and subtraction.
4.3 Add or subtract two-digit whole numbers mentally.

4.1 Make reasonable estimate when computing whole numbers.

Rounding off a number can be done to any place value of a number. We can round off to tens, hundreds, thousands, or any other power of 10.

Therefore, when we round off a number to the nearest tens, we get answers that are multiplies of 10.

Here are the rules for rounding off numbers.

Example 1: Round off 14 to the nearest 10.
- Note the number in the tens position. Here it is a 1.
- Look at the **digit** to the right of the 1.
- If this digit to the right is 5 or more, then add one to your digit in the tens position.
- If the digit to the right is not 5 or more, then call it a 0.
- So, 14 rounded off to the nearest 10 the answer is 10.

Now do the same for 28.

Example 2: Round off 28 to the nearest 10.
- Note the number in the tens position. Here it is a 2.
- Look at the digit to the right of the 2. This digit is an 8.
- The number 8 is more than 5 so we add 1 to the 2 (1 + 2 = 3). Next, write every digit after the tens position as a 0. So, 28 rounded off to the nearest 10 is 30.

This is how we write our answers. Note that for example 3 and 4 we are looking at 3-digit numbers where the digit underlined is the one we are noting.

Example 3: Round off 956 to the nearest 10.	**Example 4**: Round off 199 to the nearest 10.
Answer: 9<u>5</u>6 = 960	Answer: 1<u>9</u>9 = 200

I can buy the juice for you. It costs $145. Just give me $150.

Now try 4-digit numbers.

Example 5: Round off 1,674 to the nearest 100.	**Example 6**: Round off 5,240 to the nearest 1000.
Answer 1,<u>6</u>74 = 1,700	Answer <u>5</u>,240 = 5,000

Connect to **H**igher **E**ducation, **E**lectronic **T**ools, **A**plication and **H**elp

Let us practice 4.1.1

Let's apply the rule by rounding off to the following numbers. Complete the table below.

	Round off to the nearest ten (10)		Round off to the nearest hundred (100)		Round off to the nearest Thousand (1000)	
a.	5	g.	104	m.	995	
b.	22	h.	214	n.	1004	
c.	34	i.	449	o.	2050	
d.	49	j.	650	p.	4999	
e.	83	k.	1774	q.	7963	
f.	95	l.	2,899	r.	8421	

Use rounded numbers to estimate the answer for addition and subtraction problems.

An **estimate** is not an exact or correct answer. It is an answer which is very near to the correct answer. Knowing how to estimate a cost is an important skill used in our everyday life. For example, your parent or guardian estimates the cost of your lunch, juice and snacks to figure out how much lunch money you should get.

Steps in estimating a sum or a difference.

1. Round off the numbers to the nearest 10.
2. Use mental math to estimate the answer.

Example: 51 + 35

Rounding off	Estimated sum	Actual sum
51 → 50 35 → 40	50 + 40 —— 90	51 + 35 —— 86

Hence the estimate = 50 + 40 = 90. That means the answer is not really 90, but it is close to 90. The actual answer is 86.

Let us practice 4.1.2

Calculate the estimated and actual answers for the following problems.

Problem	Estimate	Actual	Problem	Estimate	Actual
a. 46 + 23			f. 54 – 12		
b. 54 + 16			g. 76 – 12		
c. 75 + 12			h. 35 – 18		
d. 44 + 18			i. 84 – 29		
e. 64 + 31			j. 79 – 46		

4.2 State how the properties of commutativity and associativity apply to addition and subtraction.

Communicative properties of addition and subtraction

The communicative property of **addition** states that changing the order of numbers being added does not change the sum of the numbers.

Example: 4 + 3 = 3 + 4

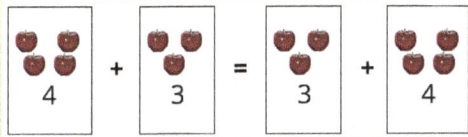

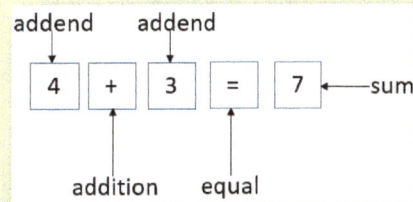

Here both sums equal 7.

The same is true in the next example where each is equals to 11.

Example: 5 + 4 +2 = 2 + 5 + 4 = 4 + 5 + 2 = 4 + 2 + 5 = 2 + 4 + 5 = 5 + 2 + 4 = 11

Summary: The order of addition does not change the sum of the numbers.

Commutative property of subtraction

Given the equation 5 - 2 = 3, the arrangement of each digit is important to the answer.

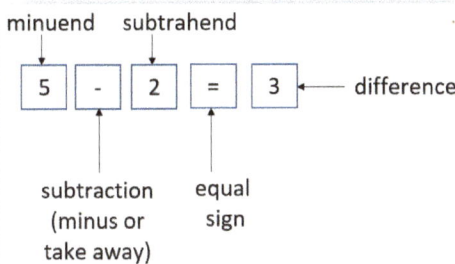

Note: An equation is a mathematical sentence that has two sides; the left-hand side (LHS) is equal to the right-hand side (RHS).

LHS = RHS

The LHS and the RHS are separated by an equal sign (=). E.g. 5 - 2 = 3.

The commutative property of **subtraction** shows that when you change around the subtrahend and the minuend the answer changes.

Example: 5 - 2 = 3 but 2 - 5 ≠ 3

≠ means not equal to

The commutative property says that the number we are subtracting can be moved or swapped from its position without changing the answer. This property is true for addition, but it is not true for subtraction. So, the commutative property shows that the order of the number is important in subtraction.

Associative property of addition and subtraction

Associative properties says that when 2 or more numbers are added or multiplied the answer is always the same and it does not matter how they are arranged or grouped. In a sum, it therefore does not matter how the numbers are set out.

Example: For addition

$$4 + 3 + 5 = 12$$
$$(4 + 3) + 5 = 7 + 5 = 12$$

$$3 + 4 + 5 = 12$$
$$4 + (3 + 5) = 4 + 8 = 12$$

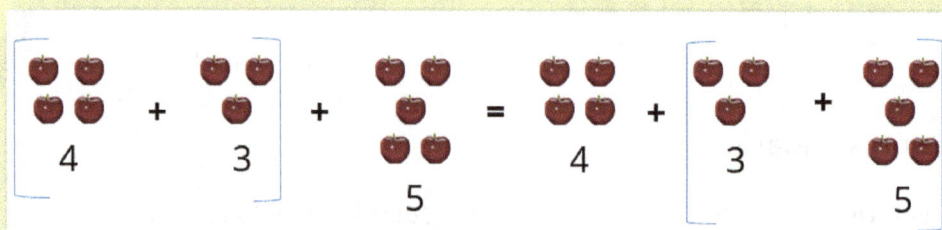

Associative property of subtraction shows that changing the order or sequence of the number subtracted changes the answer.

Example: For subtraction

$$12 - (5 - 3) = 12 - 2$$
$$= 10$$

$$(12 - 5) - 3 = 7 - 3$$
$$= 4$$

Let us practice 4.2.1

Commutative and associative properties of addition

1. Show the commutative property of addition using the following numbers:

   a. 10, 4 and 8 _____   b. 9, 7 and 12 _____

2. Show the associative property of addition using the following numbers:

   a. 12, 5 and 9 _____   b. 9, 7 and 12 _____

3. Find the missing numbers $(6 + 11) + 4 = (11 + 4) +$ _____ = _____.

4. Given $16 + 7 + n = 9 + 7 + b$, select the answer for n and b.

   a. 9, 16   b. 16, 9   c. 12, 4   d. 8, 3

5. Given $20 + (60 + P) = (60 + 20) + 5$, what is P?

   a. 8   b. 6   c. 5   d. 12

Let us practice 4.2.2

Commutative and associative properties of subtraction

1. Using a = 9 and b = 5, show that a – b ≠ b – a _____

2. Which of these is not equal to 1?

    a. 6 – 5   b. 7 – 6   c. 7 – 8   d. 4 – 3

3. Which of these is equal to 1?

    a. 8 – 7   b. 4 – 5   c. 12 – 10   d. 3 – 1

4. Which statement is true?

    a. 4 – (5 – 3) = 3 – (5 – 4)   b. 6 – 4 = 4 – 6

    c. 4 – (5 – 3) = (4 – 5) – 3   d. subtraction is commutative

4.3 Add and subtract two-digit numbers mentally.

Recall that adding **mentally** means you add numbers in your mind without writing down any steps or doing any calculation on paper or electronic devices.

When adding numbers in your mind you have to think clearly and you have to picture the image of the number in your mind.

Example 1: Here are some steps to add 42 + 29

Imagine 42 having 4 tens and 2 ones and 29 having 2 tens and 9 ones.

Therefore   ➜ 4 tens + 2 tens = 6 tens and   ➜ 2 ones + 9 ones = 1 ten plus 1 one

Together there are 6 tens + 1 ten + 1 one = 7 tens + 1 one = 71

Example 2: Here are some steps to add 23 + 4

Imagine 23 has 3 ones and 2 tens = 3 + 10 + 10. Imagine 4 has only 4 ones = 4

Therefore   ➜ 23 – 4 = 3 + 10 + 10 - 4

23 – 4 = 3 + 10 + 6

23 – 4 = 3 + 16

23 – 4 = 19

Let us practice 4.3

Let's practice mental arithmetic by doing these questions in your mind. Do not write anything when working out the answer.

a. 54 + 28	f. 37 + 17	k. 15 – 6	p. 15 – 7
b. 68 + 29	g. 35 + 29	l. 12 – 7	q. 21 – 2
c. 45 + 38	h. 49 + 37	m. 22 – 7	r. 25 – 8
d. 55 + 26	i. 79 + 16	n. 23 – 4	s. 31 – 3
e. 48 + 17	j. 84 + 17	o. 26 – 8	t. 32 – 9

Evaluation: Let us see how you did.

Learning outcome	No!	Working on it	Yes!
Did you get all the answers?	☹	😐	☺
Did you get most questions right?	☹	😐	☺
Did you retry the question(s) you got wrong?	☹	😐	☺
Were you able to correct your wrong answers?	☹	😐	☺
If not, did you seek help from others and/or review the chapter?	☹	😐	☺

Colour the face that shows how you are doing.

Focus questions

What units should I use to measure time?

What units should I use to measure lengths in my environment?

. What units should I use to measure liquids in my environment?

TERM I, UNIT 2

Chapter 5

Prior learning:

✓ Students should be familiar with the metric scale for length and mass (also called weight).

✓ Students should also be familiar with the acronym keep him down man, don't create mayhem:

Acronym	Metric Prefixes
keep	kilo
him	hecto
down	deca
man, girl, lad	metre, gram, litre
don't	deci
create	centi
mayhem	milli

Before we begin, let's see what you know.

Key vocabulary

Check the words you understand:

☐ time format

☐ estimation

☐ kilo-

☐ centi-

☐ milli-

☐ litre

☐ metre

☐ degrees Celsius

The Metric System

Prefix and base unit	Symbols for base unit metres	Symbols for base unit grams	Symbols for base unit litres	Conversions
	Length	Mass	Volume	
kilo	km	kg	kl	
hecto	hm	hg	hl	10 hecto = 1 kilo
deca	dam	dag	dal	10 deca = 1 hecto
Base unit	m	g	l	10 base unit = deca
deci	dm	dg	dl	10 deci = 1 base unit
centi	cm	cg	cl	10 centi = 1 deci
milli	mm	mg	ml	10 milli = 1 centi
Units commonly used	km, m, cm, mm	kg, g	l, ml	Note that all these measuring units are based on powers of 10.

If you look at your drink and water bottles, you see ml as the unit used.

35

5.1 Estimate, measure and record distances in metres and centimetres, in centimetres or to the nearest centimetres.
5.2 Estimate and measure straight line distances "As The Crow Flies" on a map.
5.3 Solve problems using information on a road map.
5.4 Write lengths (metres and centimetres or centimetres) in terms of a metre using decimal form.
5.5 Read and write time using the hour: minute format, e.g. 2:45 p.m.
5.6 Solve problems that involve finding time and elapsed time.
5.7 Estimate and measure capacity or volume using litres and or millilitres.
5.8 Discover that 1000 ml = 1 litre.
5.9 Identify the appropriate unit, litre, millilitre, for use in a given measurement situation.

5.1 and 5.2: Estimate, measure and record distances in metres and centimetres, in centimetres or to the nearest centimetres. Estimate and measure straight line distances "As The Crow Flies" on a map.

Look at your ruler. Do you see the cm lines numbered 0, 1, 2 ,3 and so on? Also look at the smaller lines separating the cm into 10 smaller spaces. Each of the smaller spaces is 1 mm in length.

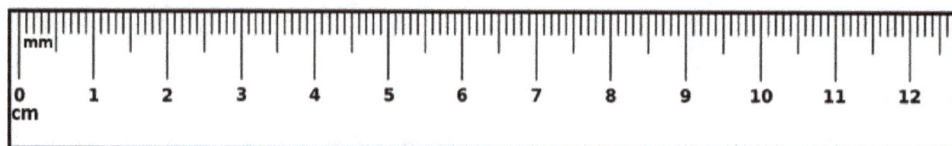

Count the number of spaces between 0 and 1. This tells that there are 10 mm in 1cm. When you measure, start from the 0 mark, not at the ruler's edge.

Let us practice 5.1 & 5.2

1. Use your ruler to draw these lines in the spaces given.

1 mm	5 mm	1 cm	1 ½ cm
5 cm		6 ½ cm	
9 cm			
1 cm			

2. Use your ruler to measure these objects in cm. Record your measurements in the table. Your ruler is much shorter than these objects and so you will need to make many measurements and add each to get the final measurement.

Object	Measurement
one side of notebook	
one side of the tabletop	
length of your pencil	
height of door	
your height	
length of your shoe	

3. Look at your ruler and familiarize yourself with what I cm looks like. Estimate (guess) the length of the straight lines to the nearest cm, then measure the lines with your ruler in cm.

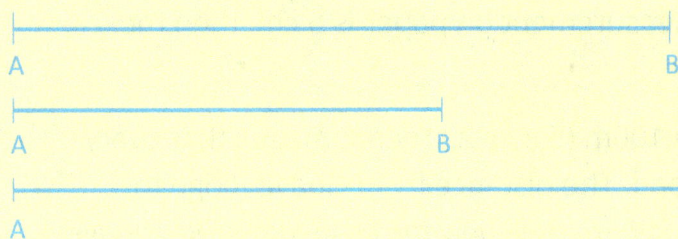

Estimate   Measurement

A ——————————————————————— B

A ——————————— B

A ——————————————————————— B

4. Estimate the length of this book to the nearest cm.

Estimate   Measurement

5. Estimate the length of this line using an appropriate metric unit. Next, using a ruler, measure and record your reading in metric units.

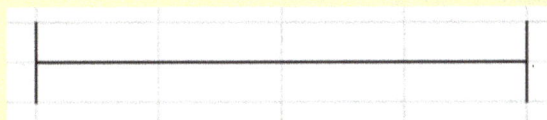

Estimate   Measurement

6. What is the difference between the real and the estimated length between lines AB and CD?

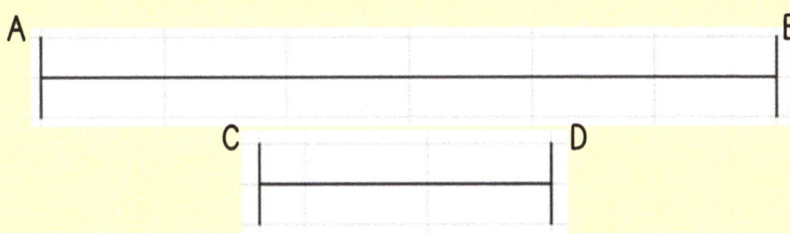

A ——————————————————————————— B

C ——————— D

Measure each line and find the difference. Compare your estimate with the real difference. How close was your estimate? _____

5.3 Solve problems using information on a road map

How can we draw Jamaica on a piece of paper? Jamaica is too large to be drawn on a piece of paper. There must be a trick to it. Yes, there is. It is called scale.

The map is called a scale drawing. The scale tells that every cm drawn on the map means a distance on the ground. For example, 1 cm:50 m means that every cm drawn on the paper in 50 m on the road. For example, on the map shown (figure 1.), suppose we want to know the distance from Guave Ridge to Riverbay Road. Use a ruler to measure the distance from Guave Ridge to Riverbay Road on the paper. If the measurement on paper is 5 cm, then the distance is equal to 5 x 50 m = 250 m.

Suppose the scale is changed to 1cm:100m (figure 2.) then it means that every cm on the paper tells 100m on the road. The distance from Guave Ridge to Riverbay Road would be 5 x 100 m = 500 m. *Note how the change in scale changes the distance*. So, the scale is **very important** on a map.

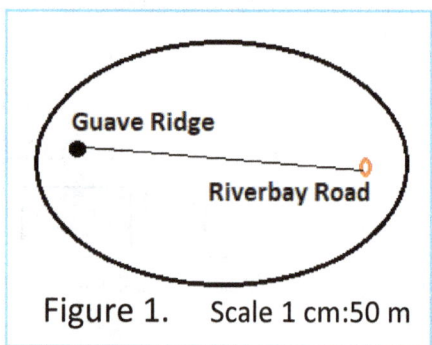

Figure 1.   Scale 1 cm:50 m

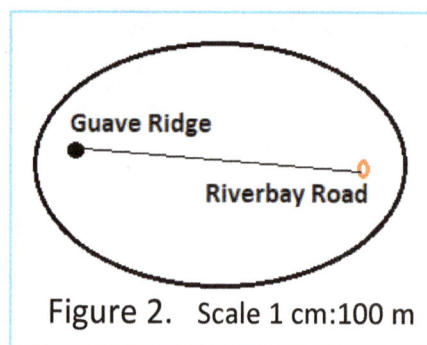

Figure 2.   Scale 1 cm:100 m

Let us practice 5.3

I.   The map below shows a community. Use the map below to answer questions (a) to (e).

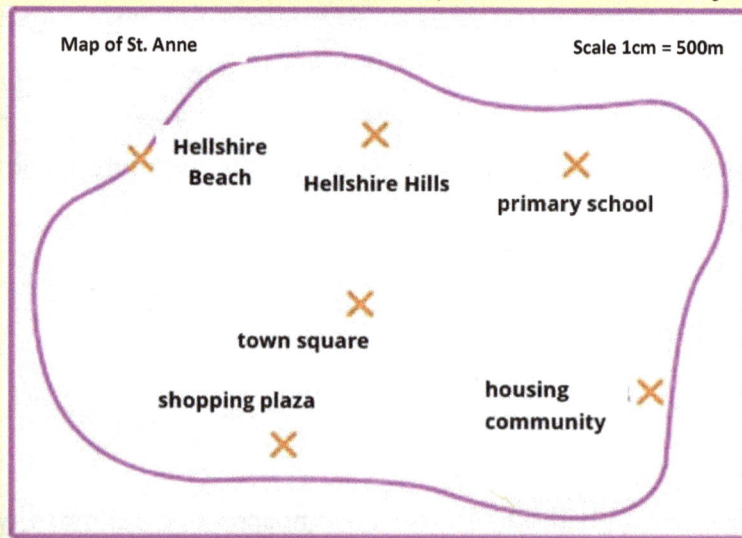

Map of St. Anne   Scale 1cm = 500m

Hellshire Beach
Hellshire Hills
primary school
town square
shopping plaza
housing community

a. How far is the housing community from the shopping plaza?
b. What is the distance from the housing community to the primary school?
c. If you travel from the housing community to Hellshire Beach and then go back home, how far did you travel?
d. A group of students walked from the primary school to the town square. How far did they travel?
e. Two girls left the housing community at the same time to go to Hellshire Beach. One girl walked to the school then to Hellshire Beach, while the other girl walked to the shopping plaza and then to Hellshire beach. Who walked the greater distance?

2. If every cm on the map below is I km on the road, how far is it from A to B?

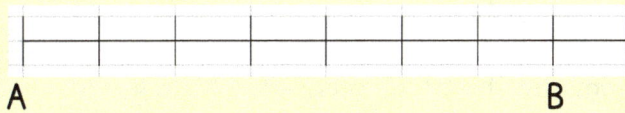

A   B

3. There are 4 shops (A, B, C and D) along the road. If on the map below each cm shows 100 m, use the length shown on the map to tell the distance between them.

A   4 cm   B   6 cm   C   8 cm   D

a. How far is A from B?   b. How far is B from C?
c. How far is C from D?   d. How far is A from D?

5.4 Write lengths (metres and centimetres or centimetres) in terms of a metre using decimal form.

Look at your ruler. Notice that 5 mm is in the middle of 1 cm. Hence 5 mm is 0.5 cm.

0.5 cm

Let us practice 5.4

1. Complete the table. Record the reading on the ruler at each letter in cm and mm using decimals.

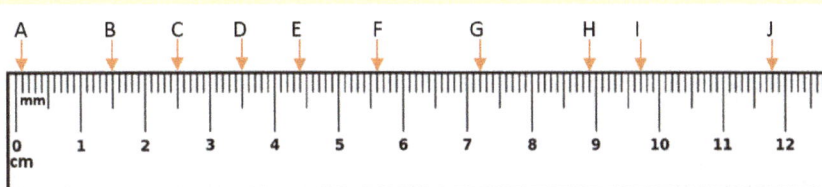

A = I mm	B = I cm 5 mm	C =	D =	E =
F =	G =	H =	I =	J =

2. Write 130 cm in metres. _____

3. Write 50 cm in metres using decimal form. _____

4. If there are 50 cm of rope, how long is this in metres? _____

5. A wire fence is 2 m and 5 cm high. What is the height of the fence in metres only?

_____

6. If the length of the book is 120 cm, write this length in metres and centimetres.

_____

7. One ruler measured 60 cm and the other measured 50 cm. What was the total length of both rulers in metres and centimetres?

_____

8. A piece of string measured 5 metres and 40 cm. Eighty cm was cut off.

   Write the length left in metres using decimals. _____

9. The long jumper landed 800 cm into the sand.

   Record this jump in metres. _____

5.5 Read and write time using the hour: minute format, e.g. 2:45 p.m

Note the analog clock shown has 2 hands. The short hand is called the hour hand and the long hand is called the minute hand.

minute hand

hour hand

The minute hand tells the number of minutes before or after the hour shown by the hour hand.

The clock reads 5 minutes past 2 o'clock.

$2:05$ AM

The digital clock operates like a computer. It uses a microchip and displays time in numbers. The clock reads 5 minutes past 2 o'clock in the morning.

Let us practice 5.5

1. Read and write the times shown on each clock.

a.

c

3.

d.

$4:10$ AM

e.

$12:35$ PM

f.

$5:45$ AM

g.	h.	i.
j.	k.	l.
m.	n.	o.
9 : 12 PM	11 : 55 AM	10 : 20

5.6 Solve problems that involve finding time and elapsed time.

Adding time

Example 1: 23 minutes 45 seconds + 45 minutes 3 seconds

min	sec
23	45
45	3
79	23

carried 1 + 79   +

1 minute (carried)

```
        1
  60 | 83
       60
       23
```
seconds

Answer = 79 minutes and 23 seconds

Example 2: 8 hours, 25 mins and 35 secs + 9 hours, 48 mins and 57 secs

hr	min	sec
8	25	35
9	48	57
18	14	32

1 + 17 = 18

1 minute (carried)
```
        1
  60 | 92
       60
       32
```
seconds

hour
```
     1
 60 | 74
      60
      14
```
minutes

Answer =18 hours, 14 minutes and 32 seconds

www.mycheetahinc.com
41
CHEETAH
Connect to Higher Education, Electronic Tools, Aplication and Help

Subtracting time
Example 1: 48 minutes 25 seconds - 15 minutes 55 seconds

min	sec
48	25
15	55
32	30

47 - 15 = 18

Convert 1 minute from the minute column to 60 seconds.

Add 60 + 25 = 85.

85 − 55 = 30 seconds.

Answer = 32 minutes and 30 seconds

Example 2: 8 hours, 35 mins and 45 secs - 3 hours, 42 mins and 50 secs

7 hours left

7 - 3 = 4

hr	min	sec
8	35	45
3	42	50
4	53	55

34 minutes left. Take one hour, convert it to 60 minutes, then add it to 34 minutes.

35 + 60 = 95

95 − 42 = 53

Take 1 minute and convert it into 60 seconds.

60 + 45 = 105.

$$\begin{array}{r} 105 \\ - \quad 50 \\ \hline 55 \end{array}$$

Answer = 4 hours, 53 minutes and 55 seconds

Multiplying time
Example 1: 16 minutes and 35 seconds x 8

4 +   16 x 8

4 +   128

= 132

min	sec
16	35
x	8
132	40

8 x 35 = 280 minutes (carried)

$$\begin{array}{r} 4 \\ 60 \overline{\smash{)}280} \\ 240 \\ \hline 40 \end{array}$$

seconds

Answer = 132 minutes and 40 seconds

Example 2: 8 hours, 45 minutes and 28 seconds x 7

7 x 8 = 56

56 + 5 = 61 hours

hr	min	sec
8	45	28
x		7
61	18	16

7 x 28 = 196 carried as minutes

$$\begin{array}{r} 3 \\ 60 \overline{\smash{)}196} \\ 180 \\ \hline 16 \end{array}$$

seconds

45 x 7 = 315

3 + 315 = 318

$$\begin{array}{r} 5 \\ 60 \overline{\smash{)}318} \\ 300 \\ \hline 18 \end{array}$$

hours carried

minutes

Answer = 61 hours, 18 minutes and 16 seconds

CHEETAH
Connect to Higher Education, Electronic Tools, Aplication and Help

Dividing time

Example 1: 14 minutes and 44 seconds ÷ 4

	min		sec
	3		41
4	14		44
	12		120
	2	4	164
			16

multiply by 60

add to seconds

x 60
120

divide by 4

4
4

Answer = 3 minutes and 41 seconds

Example 2: 16 hours, 38 minutes and 45 seconds ÷ 5

hr	min	sec
3	19	45
5) 16	38	45
15	add 60	add 180
1	38 + 60 = 98	180 + 45 = 225
x 60	19	45
60	5) 98	5) 225
	5	20
	48	25
	45	25
	3	
	x 60	
	180	

Answer = 3 hours, 19 minutes and 45 seconds

Let us practice 5.6

1. If school starts at 8:30 a.m., in how many minutes will school start?

2. Kristina started her homework at 4 o'clock and finished at the time shown. How long did it take her to finish?

3. How many minutes before it is 7 o'clock?

6 : 57 AM

4. In how many minutes will the time be 2:00 p.m.?

1 : 28 PM

5. Tom looks at his watch. The party starts at half past eight. Is Tom already late for the party?

6. The lady looked at her watch and read the time. How much time does she have, if her meeting starts at 15 minutes past eleven?

7. The train is scheduled to leave at 2:30. In how many minutes will the train leave?

1 : 45 PM

8. One bus to the city left at 6:35 p.m. A bus leaves every 40 minutes. At what time will the next bus leave?

9. The cross-country race started at 1:20 and ended at 3:45. How long did it take for the race to finish?

10. In a race, car A came in at 6:17 and car B came in 25 minutes before car A. What time did car B pass the finish line?

5.7 Estimate and measure capacity or volume using litres and or millilitres.

Litres and millilitres are used to measure the volume (capacity) of a liquid. Capacity is the amount of liquid a container can hold. The volume is the amount of liquid present in a container. When we estimate the capacity or volume, we give a value for the amount we think the container can hold or the amount we think is in the container.

Observe these measurements and read the volume in litres and millilitres.

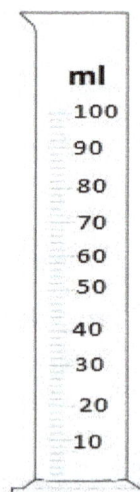

measuring cylinder

Example 1: Read the value of B on the measuring cylinder shown.

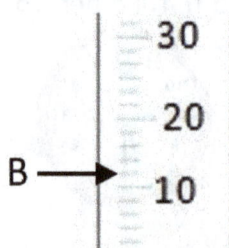

Note that B is between 10 and 20 and there are 5 smaller spaces. So, every space is 10 ml ÷ 5 = 2 ml

B is on the first line above 10 ml.

Answer: Label B is at 10 ml + 2 ml = 12 ml

Example 2: Read the value of P on the measuring cylinder shown.

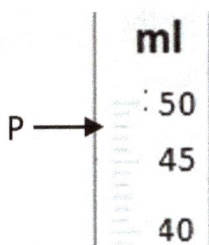

| ml | Note that P is between 50 and 45 and there are 5 smaller spaces. So, every space is 5 ml ÷ 5 = 1 ml |

P → : 50 / 45

P is on the third line above 45 ml.

40

Answer: Label P is at 45 ml + 3 ml = 48 ml

Let us practice 5.7

1. Identify the location of point R.

Label R is at

_____

ml
10
9
8
R → 7
6

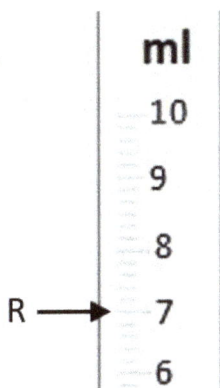

2. Read and record the approximate volume of liquid at point P.

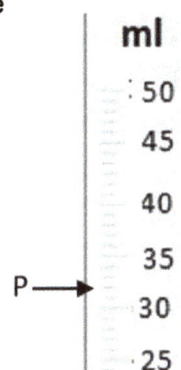

ml
: 50
45
40
35
P → 30
· 25

3. Estimate the volume of liquid in the measuring cylinder.

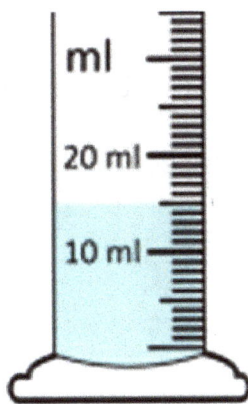

ml
20 ml
10 ml

4. Estimate the volume of liquid marked by point B.

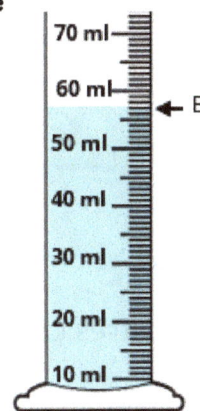

70 ml
60 ml ← B
50 ml
40 ml
30 ml
20 ml
10 ml

5. Read and record the volume of liquid in the measuring cylinder.

100 ml
80 ml
60 ml
40 ml
20 ml

6. Select the option that best represents the capacity of the measuring cylinder.

a. 80 ml   b. 20 ml
c. 100 ml   d. 110 ml

100 ml
80 ml
60 ml
40 ml
20 ml

7. Estimate the volume of medicine in the syringe.

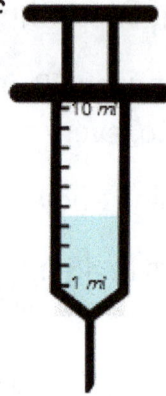

8. Select the option that best represents the capacity of the syringe.

   a. 1.0 ml
   b. 3.5 ml
   c. 10 ml
   d. 3.0 ml

9. What is the capacity of this measuring cylinder?

10. Record the volume of liquid shown in both measuring cylinders. What is the difference in volume when a stone is placed in the liquid?

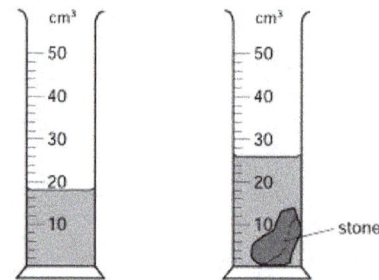

Difference in volume = _____

5.8 Discover that 1000 ml = 1 litre.

Millilitres can be converted to litres.

> 1000 millilitres = 1 litre

Example: 500 ml of orange juice was poured into a jug which has 800 ml of water.

i. How much juice is in the container in L?

    500 ml + 800 ml = 1,300 ml.
    1300 ml = 1 litre and 300 ml.

ii. If the capacity of the jug is 2 L, how much space is left in the jug?

    2000 ml – 1300 ml = 700 ml

Let us practice 5.8

1. Circle the measuring cylinder with a volume greater than one litre.

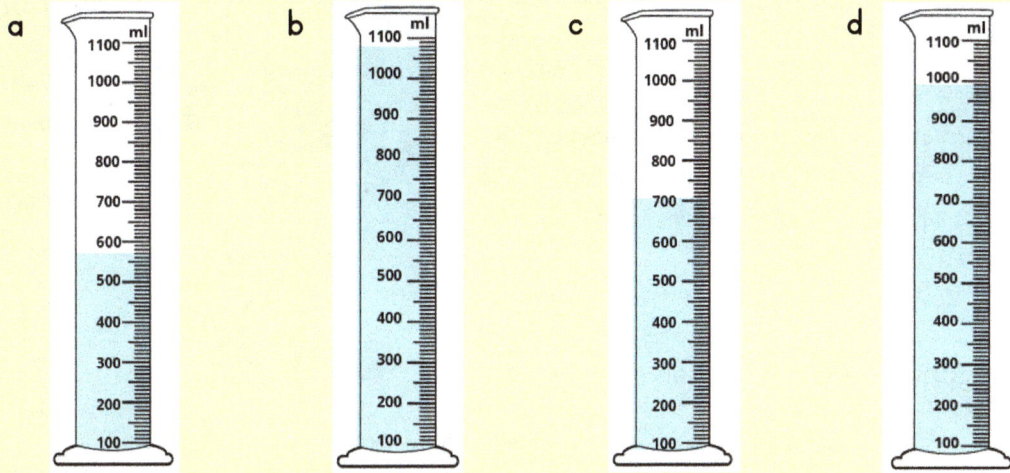

a   b   c   d

2. Four (4) small cups of water in cup A were poured into cup B. What volume of water would be in cup B?

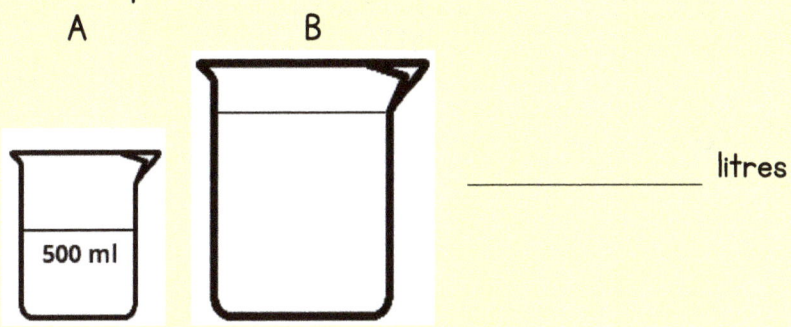

A   B

500 ml

_____ litres

3. Two cups each with 250 ml were taken out. What volume is left? Give your answer in litre and ml.

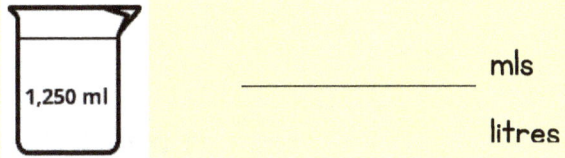

1,250 ml

_____ mls

litres

Evaluation: Let us see how you did.

Learning outcome	No!	Working on it	Yes!
Did you get all the answers?	☹	😐	🙂
Did you get most questions right?	☹	😐	🙂
Did you retry the question(s) you got wrong?	☹	😐	🙂
Were you able to correct your wrong answers?	☹	😐	🙂
If not, did you seek help from others and/or review the chapter?	☹	😐	🙂

Colour the face that shows how you are doing.

What units should I use to measure the mass of objects in my environment?
What units should I use to measure temperature in my environment?

Before we begin, let's see what you know.

Prior learning:

✓ Explain the relationships between the units having the prefixes deci-, centi-, milli-and kilo- and the main units.

✓ Associate units of measurement and instruments to appropriate items.

Key vocabulary

Check the words you understand:

☐ Celsius
☐ degree
☐ gram
☐ kilogram
☐ mass
☐ temperature
☐ thermometer

1. Recall mass conversion from the table below.

Metric unit	Symbol	Conversion rates
kilogram	kg	
hectogram	hg	10 hg = 1 kg
decagram	dag	10 dag = 1 hg
gram	g	10 g = 1 dag
decigram	dg	10 dg = 1 g
centigram	cg	10 cg = 1 dg
milligram	mg	10 mg = 1cg

To move from milli- to centi- to deci- and up the table we divide each by 10.

2. Draw lines to match the following prefixes to their value in the metric conversion.

kilo-   divide by 10

deci-   divide by 1000 or multiply by $\frac{1}{1000}$

centi-   multiply by 1000

milli-   divide by 100

3. Draw lines to match the instrument used to measure the following metric units. One has been done for you.

kg   thermometer

ᵒC   measuring cup

m   scale balance

l   ruler

cm   tape measure

If you know these, you should be able to learn what comes next.

6.1 Estimate and measure mass using gram or kilogram or kilogram and gram.
6.2 Read a scale shown in a measurement situation using kilograms and/or grams.
6.3 Discover that 1000 kg = 1 tonne.   6.4 Estimate and measure temperature in degrees Celsius.
6.5 Tell the difference between two temperatures both above zero.
6.6 Tell the temperature which is a given number of degrees warmer or cooler than a given temperature.

6.1 Estimate and measure mass using gram or kilogram or kilogram and gram.

People who sell goods such as shop keepers and market sellers can often look at their goods and tell you the **mass** (weight). They are usually very close, if not precise, in their estimates. To estimate mass accurately, some practice is needed to get an idea of what the actual mass looks and feels like for that product. While shopping at a supermarket, you can look at and lift groceries where the mass is measured and labelled in kg.

Scales are instruments used to measure mass. How they look depends on what they are used to measure.

You can also practice with your groceries at home. Look at the parcels of rice, flour, sugar and cornmeal and estimate their mass in kg. Then, lift them and estimate their mass, then look at the mass on the label of the bags. Compare the actual mass to your estimates.

Write your estimate and actual mass for the groceries below using kg as your unit.

Grocery	Estimated mass from looking at object	Estimated mass from feeling the weight of the object	Actual mass on the bag's label
rice			
flour			
sugar			
cornmeal			

*Be sure to note if some of the groceries were used from each bag before using them for this activity.

You will also find that the feel of different mass will be different based on the shape of the object. Try these examples.

1. To estimate a 500 g mass, fill a 500 ml plastic bottle of water and lift it several times to get the feel of the mass (weight).

2. To estimate a 1 kg mass, fill a 1 litre plastic bottle of water and lift it several times. to get the feel of the mass (mass).

3. You may also go to the gymnasium and look at the mass of some objects and test them.

4. Get permission at home to use the kitchen scale safely and estimate the mass of some household items, then weigh them.

Let us practice 6.1

1. Think of these objects and tick the unit of measurement that you think would be suitable to measure the object.

a.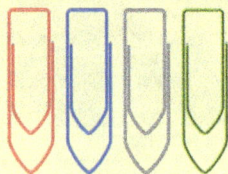
☐ grams
☐ kilograms

b.
☐ grams
☐ kilograms

c.
☐ grams
☐ kilograms

d.
☐ grams
☐ kilograms

e.
☐ grams
☐ kilograms

f.
☐ grams
☐ kilograms

g.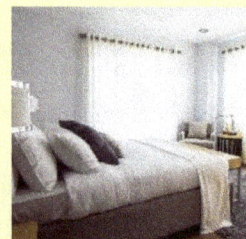
☐ grams
☐ kilograms

6.2 Read a scale shown in a measurement situation using

Let us practice 6.2

1. Read from the scales below and record the mass of each item in grams or kilograms.

a.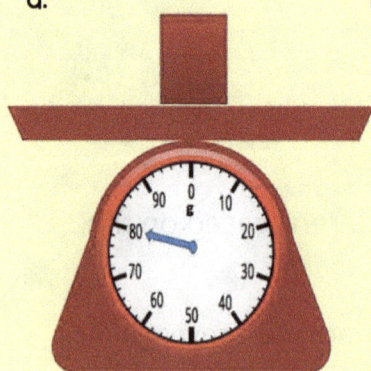

b.

c.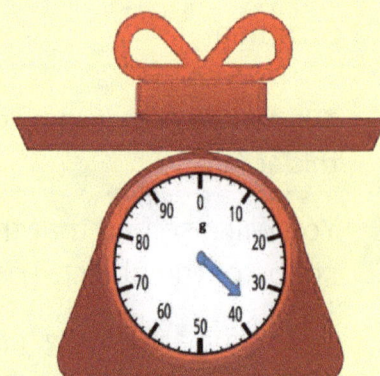

_____   _____   _____

d.

_____

e.

_____

f.

_____

g.

_____

h.

_____

i.

_____

2. Practise by converting these measurements.

a.	2 kg = g	i.	17 kg = g	p.	8000 g = kg		
b.	3 kg = g	j.	20 kg = g	q.	9000 g = kg		
c.	5 kg = g	k.	24 kg = g	r.	10,000 g = kg		
d.	8 kg = g			s.	12000 g = kg		
e.	10 kg = g	l.	1,000 g = kg	t.	15000 g = kg		
f.	11 kg = g	m.	2000 g = kg	u.	19000 g = kg		
g.	13 kg = g	n.	4000 g = kg	v.	23,000 g = kg		
h.	16 kg = g	o.	7000 g = kg				

CHEETAH™
Connect to **H**igher **E**ducation, **E**lectronic **T**ools, **A**plication and **H**elp

3. Practise by converting these measurements.

a.	1 kg	150 g=	g	l.	1,200 g =	kg		g
b.	2 kg	200 g=	g	m.	1,000 g =	kg		g
c.	4 kg	380 g=	g	n.	1,234 g =	kg		g
d.	6 kg	500 g=	g	o.	2,500 g =	kg		g
e.	7 kg	250 g=	g	p.	3,550 g =	kg		g
f.	13 kg	600 g=	g	q.	5,000 g =	kg		g
g.	15 kg	151 g=	g	r.	6,439 g =	kg		g
h.	16 kg	234 g=	g	s.	7,103 g =	kg		g
i.	18 kg	974 g=	g	t.	8,002 g =	kg		g
j.	21 kg	356 g=	g	u.	9,134 g =	kg		g
k.	24 kg	778 g=	g	v.	10,001 g =	kg		g

4.   Draw the pointer on each scale to show the mass given.

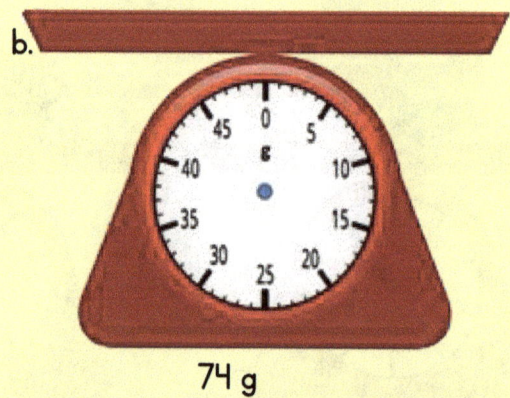

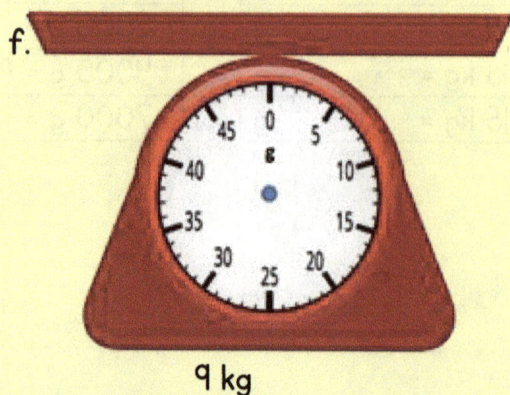

a.

28 g

b.

74 g

c.

2 kg

d.

17 g

e.

26 g

f.

9 kg

6.3 Convert these masses to kilograms; Discover that 1000 kg = 1 tonne.

The mass of an object is commonly measured in grams (g), kilograms (kg) and sometimes tonnes (ton).

Note that 1000 kg = 1 metric tonne and 1 kg = $\frac{1}{1000}$ ton

Let us practice 6.3

1. The truck is labelled to carry no more than 6 tonnes. Which of the following loads can it carry safely?

    a. 2000 kg + 3000 kg + 2000 kg   b. 4000 kg + 3000 kg
    c. 1000 kg + 3000 kg + 500 kg   d. 6000 kg + 500 kg + 4 kg

2. A truck with a limit of 8 tonnes had a mass of 9 ½ tonnes. By how much was it over its limit?

    a. 17 ½ tonnes   b. 72 tonnes
    c. 1 ½ tonnes   d. 3 ½ tonnes

3. The stone has a mass of 4 tonnes. What is the weight of the stone in kg?

4. A bridge was built to carry motor vehicles which weigh less than 5000 kg. Should a truck loaded with stones weighing 6 tonnes be allowed to cross the bridge? Explain why?

5. What is the mass of a truck weighing 8.5 tonnes in kg and g?

6.4 Estimate and measure temperature in degrees Celsius.

A **thermometer** is an instrument used to tell how hot or cold the atmosphere is. In Jamaica, we use the **Celsius** scale to measure atmospheric **temperature**.

Each thermometer may use a different multiplier to show the temperature. Some thermometers count in 10s, example 10, 20, 30 and so on. Other thermometers count in 2s or 20s. Some thermometers have to be read, while others have a digital screen similar to a digital watch from which the temperature is displayed. Note that we sometimes have to estimate the temperature from the thermometers we read.

Each thermometer measures the temperature in **degrees** Celsius (°C); practice reading and writing your estimated temperature in the space given.

Thermometers are instruments used to measure temperature. They measure temperature in degrees Celsius (°C), Fahrenheit (°F) or kelvin (°K).

40°C

25°C

Let us practice 6.4

1. What is the reading on the thermometer?

   a. 32°C   c. 22°C

   b. 26°C   d. 18°C

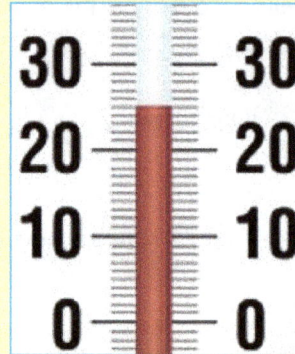

2. Read and record the temperatures (°C) at the two towns.

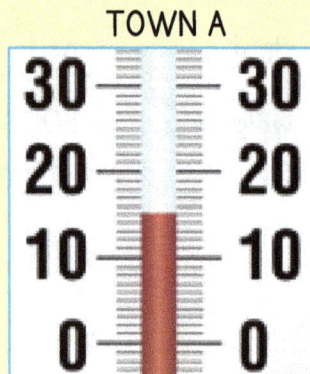

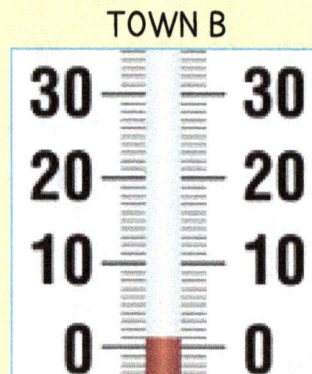

   TOWN A   TOWN B

   _____   _____

   Which statement is true?

       i. Town A is warmer than Town B.

       ii. Town B is warmer than Town A

       iii. Both towns have temperatures greater than 0 °C.

       iv. Both towns have very hot temperatures.

   a. I only   b. II and IV   c. II and III   d. I and III

3. The air conditioner in the office read 31 °C. The secretary changed it to 24 °C. Which of these statements is true?

   a. The office became warmer by 7 °C.

   b. The office became colder by 7 °C.

   c. The office became colder by 55 °C.

   d. The office became warmer by 55 °C.

CHEETAH
Connect to Higher Education, Electronic Tools, Aplication and Help

4. What is the reading on the thermometer in degree Celcius?

   a. 24 °C   c. 23 °C
   b. 22 °C   d. 20 °C

5. Read this thermometer.

   a. Greater than 39 °C
   b. Less than 39 °C
   c. Greater than 40 °C
   d. Exactly 40 °C

6. Aiden has a big spelling test at school this morning. When his mother comes to wake him, he says he had=s a fever and cannot go to school because he is sick. Aiden's normal body temperature is between 36 °C and 37 °C. Read the thermometer and choose the answer that best explains how Aiden is feeling.

   a. Aiden's temperature is in the normal range, so he is feeling ill from a fever.
   b. Aiden's temperature is not in the normal range, so he is feeling ill from a fever.
   c. Aiden's temperature is not in the normal range, and he does not want to go to school
   d. Aiden's temperature is in the normal range, and he does not want to go to school.

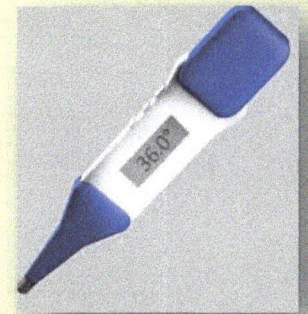

6.5 Tell the difference between two temperatures both above zero.

The difference between temperatures is calculated by subtracting the temperature readings.

Example: If the temperature inside a house is 26 °C and the temperature outside is 32 °C, by how much is the temperature different? Underline the letter for the correct answer.

Temperature difference = 32 °C - 26 °C = 6 °C

   a. 58 °C   b. 6 °C   c. 5 °C   d. no difference

Example: By how many degrees is place P cooler than place Q? Calculate their temperature difference and select your answer below.

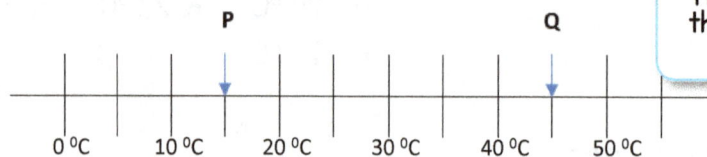

P   Q

Hint: Read and record the temperatures at P and Q.

0 °C   10 °C   20 °C   30 °C   40 °C   50 °C

a. 30 °C   b. 20 °C   c. 15 °C   d. 45 °C

Let us practice 6.5

Complete the table. Calculate the temperature difference.

Question	Start temperature	End temperature	Change in temperature
1.	16 °C	42 °C	
2.	56 °C	10 °C	
3.	18 °C	78 °C	
4.	71 °C	6 °C	
5.	55 °C	5 °C	
6.	2 °C	8 °C	

6.6 Tell the temperature which a given number of degrees is warmer or cooler than a given temperature.

When we compare two thermometers, the warmer temperature is higher in degrees Celsius (°C).

Note that the larger the number of degrees Celsius (°C), the warmer the temperature will be. So, 32 °C is warmer than 26 °C.

Example: State which thermometer shows warmer temperature and by how much.

30°C   30°C

20   20

10   10

0   0

Thermometer 1 (T$_1$)   Thermometer 2 (T$_2$)

$T_1 = 24$ °C

$T_2 = 21$ °C

24 °C – 21 °C = 3 °C

T_1 is warmer by 3 °C

Let us practice 6.6

1. Look at the temperature on both thermometers. Record the temperatures, tell which is colder and by how much.

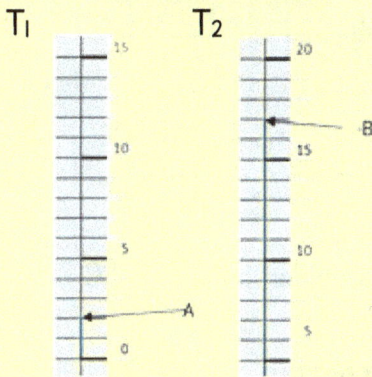

T₁   T₂

A = _____ °C   B = _____ °C

Difference = _____ °C - _____ °C

= _____ °C

_____ is colder by _____ °C

2. Look at the temperature on both thermometers. Record the temperatures, say which is warmer and by how much.

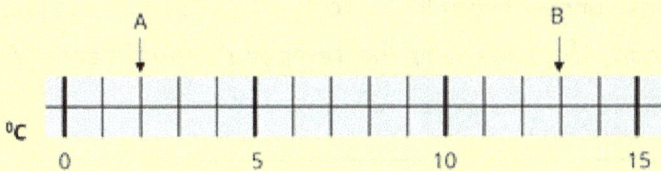

56 °C	89 °C
A	B

A = _____

B = _____

Difference: _____ °C - _____ °C = _____

= _____ °C

_____ is warmer by _____ °C

3. Record the temperatures at two places A and B. Say which is colder and by how much.

A = _____ °C   B = _____ °C

Difference = _____ °C - _____ °C

= _____ °C

_____ is colder by _____ °C

4. Which place is warmer, A or B and by how many °C?

Kingston (A): 27 °C   Athens (B): 9 °C

   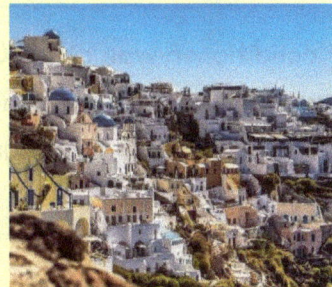

A = _____ °C   Difference = _____ °C - _____ °C

B = _____ °C   = _____ °C

_____ is warmer by _____°C

CHEETAH
Connect to Higher Education, Electronic Tools, Aplication and Help

5. Which location, A or B, is colder and by how many degrees °C?

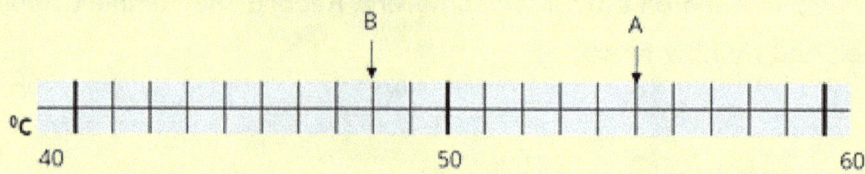

B   A

°C 40   50   60

A = _____ °C   Difference = _____ °C - _____ °C

B = _____ °C   = _____ °C

                                _____ is colder by _____ °C

6. Which reading, P or Q, on the thermometer shows a hotter temperature. Read and record the temperature difference.

Write your answers here:

_____

_____

°C
35
30 ← P
25
20 ← Q
15
10
5

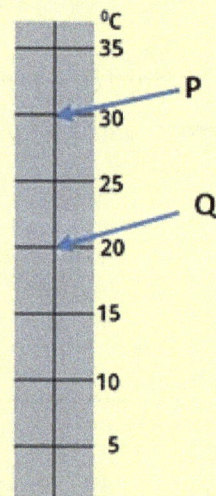

7. The air conditioner in the office was turned from 18 °C to 25 °C. Was the office made cooler or warmer? By how many degrees was the temperature different?

_____

8. The air conditioner in the office was turned from 24 °C to 18 °C. Will the workers put their sweater on or take them off. Give a reason for your answer.

_____

9. The air conditioner was at 21 °C. The office attendant turned off the air conditioner and opened the window. Do you expect the office to get colder or warmer. Give a reason for your answer.

_____

10. The temperature of the room was 23 °C. The temperature was changed and warmed by 8 °C. What was the new temperature?

_____

Evaluation: Let us see how you did.

Learning outcome	No!	Working on it	Yes!
Did you get all the answers?	☹	😐	🙂
Did you get most questions right?	☹	😐	🙂
Did you retry the question(s) you got wrong?	☹	😐	🙂
Were you able to correct your wrong answers?	☹	😐	🙂
If not, did you seek help from others and/or review the chapter?	☹	😐	🙂

Colour the face that shows how you are doing.

CHEETAH
Connect to Higher Education, Electronic Tools, Aplication and Help

Focus question

TERM I, UNIT

How do I record and compute the various units of measurement?

Chapter

Before we begin, let's see what you know.

Prior learning:
✓ Associate units of measurement with their symbols.
✓ Associate an item to be measured with its appropriate unit.

Key vocabulary

Check the words you understand:

☐ estimation
☐ grammes
☐ kilo-
☐ centi-
☐ milli-
☐ litre
☐ metre
☐ time format

1. Complete the following table.

Measurement	Base units	Units of measurement
mass (weight)	_____ or _____	_____ or kg
_____	metre or _____	_____ or cm
volume or capacity	_____	_____
_____	seconds, _____ or _____	_____, mins or hr
_____	degree Celsius	_____

If you know these, you should be able to learn what comes next.

🎯 7.1 Discuss the general meaning of the prefixes deci-, centi-, milli-, kilo-.
7.2 Explain the relationships between the units having the prefixes deci-, centi-, milli-and kilo- and the main units; gram, metre and litre.
7.3 Convert one unit of measurement to another (length: kilometres and metres).
7.4 Explain the relationships among units of time.
7.5 Convert one unit of measurement to another (time: hours, minutes and seconds).
7.6 Convert one unit of measurement to another (capacity: millilitres and litres).
7.7 Convert one unit of measurement to another (mass: kilograms and grams).

7.1 Discuss the general meaning of the prefixes deci-, centi-, milli-, kilo-.

Metric units can be converted, as they are all based on powers of ten.

Unit / prefix	Meaning
kilo-	1000 times larger than base unit
metre gram litre	**Base units** metre – length and distances gram – mass (weight) litre – capacity of liquids; volume of solids and liquids
deci-	10 times smaller than base unit
centi-	100 times smaller than base unit
milli-	1000 times smaller than base unit

7.2 Explain the relationships between the units having the prefixes deci-, centi-, milli-and kilo- and the main units; gram, metre and litre.

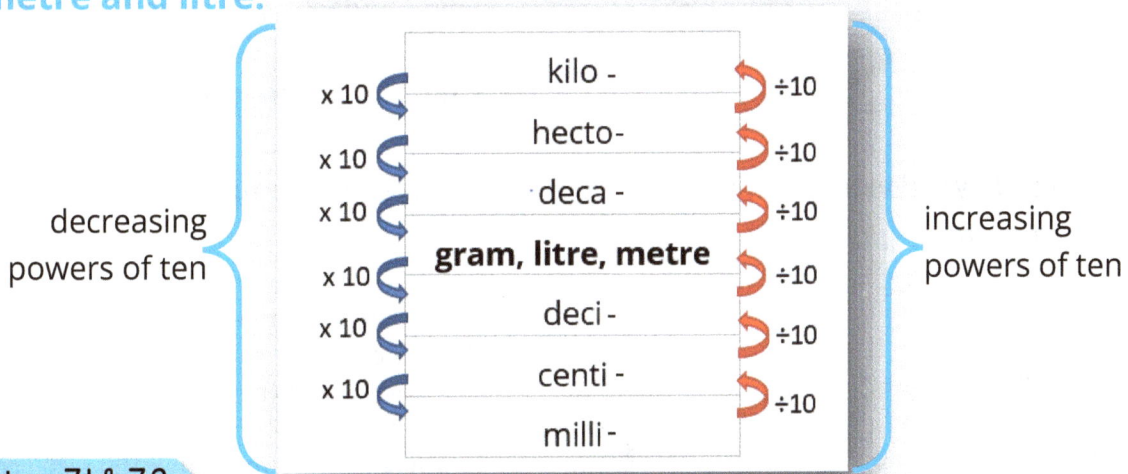

decreasing powers of ten

kilo -
x 10 ÷10
hecto-
x 10 ÷10
deca -
x 10 ÷10
gram, litre, metre
x 10 ÷10
deci -
x 10 ÷10
centi -
x 10 ÷10
milli -

increasing powers of ten

Let us practice 7.1 & 7.2

How many times larger or smaller is the prefix to the base unit?

Prefixes	Relationship between prefixes and base unit
deca-	_____ times larger than base unit
hecto-	_____ times larger than base unit
kilo-	_____ times larger than base unit
milli-	_____ times smaller than base unit
centi-	_____ times smaller than base unit
deci-	_____ times smaller than base unit

7.3 Convert one unit of measurement to another (length: kilometres and metres).

When using the prefixes for any base unit, kilo is the largest and milli is the smallest.

When converting from big units to smaller units we multiply. When converting from small units to bigger units we divide.

How do we decide what to divide or multiply by when converting from one metric unit to another? Cover the one that you get and see how many jumps are needed to get to the unit you want.

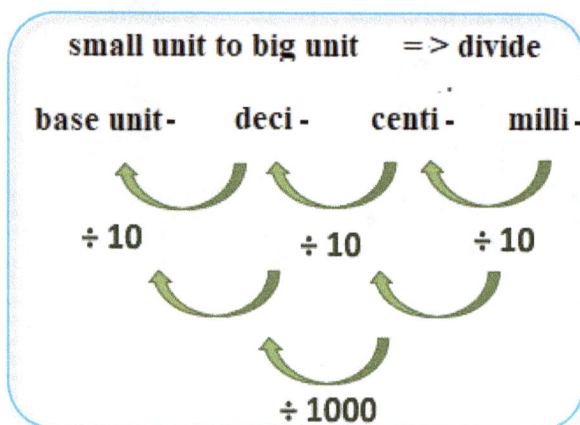

So, what do we divide with or multiply by? Let us take for example converting from km to m.

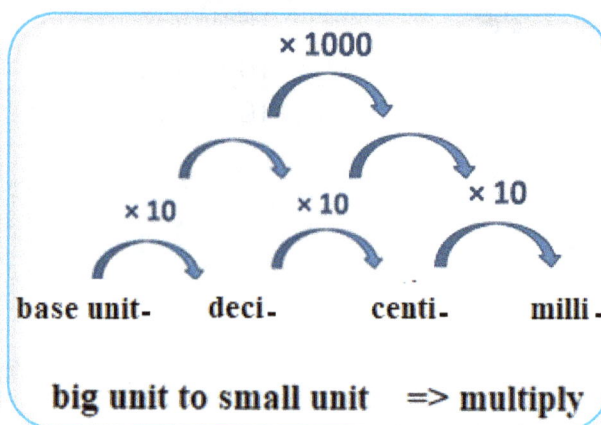

	Converting metric units
km to m	Multiply by 1000 **Example:** 2 km to m= 2 km x 1000 = 2,000 m
m to cm	Multiply by 100 **Example:** 3 m to cm = 3 m x 100 = 300 cm
m to mm	Multiply by 1000 **Example:** 4 m to mm = 4 m x 1000 = 4,000 mm
cm to mm	Multiply by 10 **Example**: 5 cm to mm = 5 cm x 10 = 50 mm

So, 2 km would be multiplied by 1000 to give the equivalent measurement in m of 2000 m. Use the table above to practise converting from bigger units to smaller units through multiplication.

Now let us convert from m to km. So, 48,000 m would be divided by 1000 to give the equivalent measurement in km of 48 km. Use the table below to practice converting from smaller units to bigger units through division.

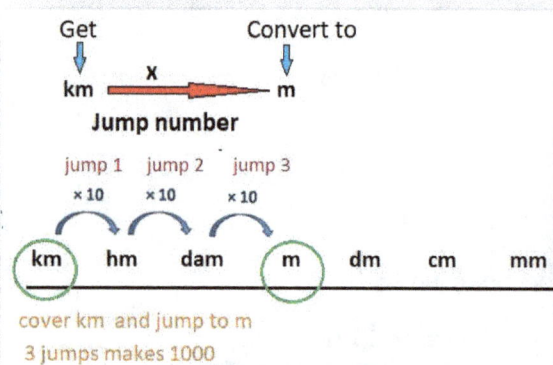

	Converting metric units
mm to cm	Divide by 10 **Example:** 70 mm to cm = 70 mm ÷ 10 = 7 cm
mm to m	Divide by 10 **Example:** 12,000 mm to m = 12,000 mm ÷ 1000 = 12 m
cm to m	Divide by 100 **Example:** 8,000 cm to m = 8,000 cm ÷ 100 = 80 m
m to km	Divide by 1000 **Example:** 48,000 m to km = 48,000 m ÷ 1000 = 48 km

Let us practice 7.3

1. Convert the following metric units for lengths and distances.

cm to mm		mm to cm	
i. 8 cm to mm		xvi. 20 mm to cm	
ii. 10 cm to mm		xvii. 60 mm to cm	
iii. 46 cm to mm		xviii. 120 mm to cm	
iv. 351 cm to mm		xiv. 200 mm to cm	
v. 728 cm to mm		xx. 980 mm to cm	
vi. 2 m to cm		xxi. 100 cm to m	
vii. 5 m to cm		xxii. 300 cm to m	
viii. 26 m to cm		xxiii. 700 cm to m	
ix. 36 m to cm		xxiv. 1000 cm to m	
x. 136 m to cm		xxv. 1,200 cm to m	
xi. 1 km to m		xxvi. 2000 m to km	
xii. 2 km to m		xxvii. 3000 m to km	
xiii. 3 km to m		xxviii. 5000 m to km	
xiv. 6 km to m		xxiv. 2000 m to km	
xv. 9 km to m		xxx. 56,000 m to km	

2. A student walked from point A to point B as shown on the diagram. How many metres did she student walk?

   _____ m

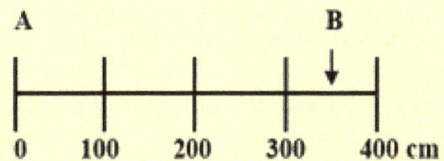

3. A planter walked around his garden. What was the distance walked in metres?

   Distance = _____ m

4. Read the ruler and record the distance from A to B in metres.

   _____ m

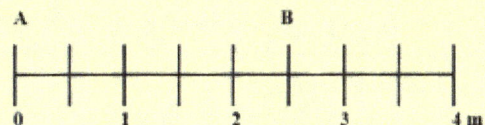

5. Trace the outline of the book. The total length around the book is I m. The length of the long side is 30 cm. Write the length of the long side beside each long side. How long is the length of one of the short sides in cm?

_____ cm

6. If Nicholas made 10 steps and one step was 50 cm, how many metres did he walk?

_____

_____

7. Keisha has 2 m of fishing line and John has 1 ½ m of fishing line. How many centimetres of fishing line do they have altogether?

_____

_____

8. Both Shameila and Ashawn measured their pencils. Shameila's pencil was 8 cm longer than Ashawn's. If Ashawn's pencil was 6 cm, what is the length of Shameila's pencil in mm?

_____

_____

9. If Kelvin walked 2 m and Amia walked 50 m more than Kelvin, how many cm did they walk altogether?

_____

_____

10. The bus stopped 3 metres from the stop sign. How many cm away did it stop?

_____

_____

11. There are 5 cars parked along the school fence. If each car is 2 ½ m long and there is ½ m between each car. How much of the roadway does the line of cars occupy?

_____

_____

7.4 Convert one unit of measurement to another (capacity: millilitres and litres).

Let us change small units to big units and vice versa.

1000 ml = 1 L

Converting metric units for volume and capacity	
ml to L	Divide by 1000 **Example:** 2000 ml to L = 2000ml ÷ 1000 = 2 L
L to ml	Multiply by 1000 **Example:** 3 L to ml = 3 L x 1000 = 3000 ml

Let us practice 7.4

1. Convert to L

a. 2,000 ml =	c. 1,000 ml =	e. 8,500 ml =	g. 8,000 ml =	i. 10,500 ml =
b. 15,000 ml =	d. 6,000 ml =	f. 1,500 ml =	h. 25,000 ml =	

2. Convert to ml.

a. 2 L =	c. 10 L =	e. 8 L =	g. 7 ½ L =	i. 120 L =
b. 7 L =	d. 3 ½ L =	f. 50 L =	h. 8½ L =	

3. Convert to litres and ml.

a. 1 ½ L =	c. 5 ½ L =	e. 8 ½ L =
b. 2 ½ L =	d. 7 L =	f. 10 ½ L =

4. Convert to litres and ml.

a. 2,355 ml =	c. 7,500 ml =	e. 8,364 ml =
b. 6,001 ml =	d. 5,356 ml =	f. 38,545 ml =

5. If there are 2 L of water in the pan boiling and 20 ml leaks out of the pan every 1 minute, how much water will be in the pan after 10 minutes?

_____

_____

7.5 Convert one unit of measurement to another (mass: kilograms and grams).

Let us change small units to big units and vice versa.

Converting metric units for mass or weight	
mg to g	Divide by 1000 **Example:** 2000 mg to g = 2000 ÷ 1000 = 2 g
g to kg	Divide by 1000 **Example:** 2000 g to kg = 2000 ÷ 1000 = 2 kg
g to mg	Multiply by 1000 **Example:** 3 g to mg = 3 g x 1000 = 3000 mg
kg to g	Multiply by 1000 **Example:** 4 kg to g = 4 kg x 1000 = 4,000 g

CHEETAH
Connect to Higher Education, Electronic Tools, Aplication and Help

Let us practice 7.5

1. Convert to mg.

a. 1 g =	d. 2 ½ g =	g. 5 ½ g =
b. 6 g =	e. 7 g =	h. 8 ½ g =
c. 16 g =	f. 25 g =	i. 220 g =

2. Convert to g

a. 1,000 mg =	d. 2,000 mg =	g. 5,000 mg =
b. 7, 000 mg =	e. 9,500 mg =	h. 15,000 mg =
c. 1,500 mg =	f. 3,500 mg =	i. 4,500 mg =

3. Convert to kg

a. 1,000 g =	d. 3,000 g =	g. 6,000 g =
b. 7, 000 g =	e. 8,500 g =	h. 12,000 g =
c. 1,500 g =	f. 2,500 g =	i. 5,500 g =

4. Convert to g and mg

a. 2 ½ g =	d. 3 ½ g =	g. 8 ½ g =
b. 13 ½ g =	e. 6 ½ g =	h. 17 ½ g =
c. 260 mg =	f. 4,900 mg =	i. 7,500 mg =

5. Convert to kg and g

a. 3,401 g =	c. 4,600 g =	e. 8,294 g =
b. 5,755 g =	d. 5,660 g =	f. 10,545 g =

6. Convert to g

a. 3 kg and 1g = _____ g	d. 5 kg and 6 g = _____ g
b. 6 kg and 755 g = _____ g	e. 14 kg and 609 g = _____ g
c. 7 kg and 294 g = _____ g	f. 8 kg and 55 g = _____ g

7. A man was carrying a load of 3 ½ kg. What reading would the load give on a scale which only measured in grams? _____ g

8. Amy bought a bag of corn which was marked 12 ½ kg of the bag. When she got home, she weighed the bag and found that it was 500 g short. How much did it really weigh?

_____

_____

9. A 5 kg bag of potatoes weighed ½ kg more than was written on the bag. What was the real mass of the bag of potatoes in grams?

_____

_____

10. The total mass of 4 students' lunch was 7 ½ kg. If 3 of the masses were 1 ½ kg, 2 kg and 2 ½ kg, what was the mass of the 4th student's lunch measured in grams?

_____

_____

7.6 Explain the relationships among units of time.

The units of time are interrelated.

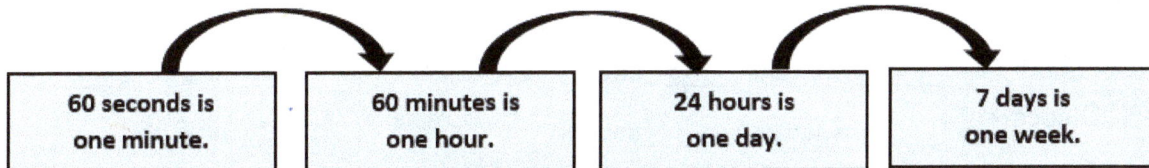

60 seconds is one minute.	60 minutes is one hour.	24 hours is one day.	7 days is one week.

Let us change big units to small units.

Converting metric units	
mins to secs	Multiply by 60 **Example:** 2 mins = 2 x 60 = 120 secs
hrs to mins	Multiply by 60 **Example:** 3 hrs = 3 x 60 = 180 mins
days to hrs	Multiply by 24 **Example:** 2 days = 2 x 24 = 48 hrs
weeks to days	Multiply by 7 **Example:** 5 weeks = 5 x 7 = 35 days

Let us change small units to big units.

Converting metric units	
secs to mins	Divide by 60 **Example**: 120 secs = 120 ÷ 60 = 2 mins
mins to hrs	Divide by 60 **Example:** 180 mins = 180 ÷ 60 = 3 hrs.
hrs to days	Divide by 24 **Example:** 48 hrs = 48 ÷ 24 = 2 days
days to weeks	Divide by 7 **Example:** 28 days = 28 ÷ 7 = 4 weeks

7.7 Convert one unit of measurement to another (time: hours, minutes and seconds).

Convert 376 minutes to hours and minutes.

$$
\begin{array}{r}
6 \leftarrow \text{hours} \\
60\overline{)376} \\
-360 \\
\hline
16 \leftarrow \text{minutes}
\end{array}
$$

Answer = 6 hrs. and 16 mins

Convert 420 seconds to minutes and seconds

$$
\begin{array}{r}
7 \leftarrow \text{minutes} \\
60\overline{)450} \\
-420 \\
\hline
30 \leftarrow \text{seconds}
\end{array}
$$

Answer = 7 mins and 30 sec

Let us practice 7.6 & 7.7

1. Convert one unit of measurement of time to another.

a. 60 secs =	mins.	m. 60 mins =	hrs.
b. 120 secs =	mins.	n. 240 mins =	hrs.
c. 180 secs =	mins.	o. 360 mins =	hrs.
d. 240 secs =	mins.	p. 420 mins =	hrs.
e. 480 secs =	mins.	q. 720 mins =	hrs.
f. 600 secs =	mins.	r. 1800 mins =	hrs.
g. 1200 secs =	mins.	s. 2400 mins =	hrs.
h. 365 secs =	_____mins and _____secs	t. 245 mins =	_____hrs. and _____mins
i. 400 secs =	_____mins and _____secs	u. 400 mins =	_____hrs. and _____mins
j. 500 secs =	_____mins and _____secs	v. 500 mins =	_____hrs. and _____mins
k. 600 secs =	_____mins and _____secs	w. 560 mins =	_____hrs. and _____mins
l. 1220 secs =	_____mins and _____secs	x. 720 mins =	_____hrs. and _____mins

2. Draw lines to match equal times.

300 seconds	135 seconds
2 minutes and 15 seconds	120 minutes
2 hours	14 days
72 hours	5 minutes
2 weeks	3 days

3. Paul's numeracy test lasted 2 ½ hours. How many minutes did he have to complete this test? _____

4. The bus took 300 minutes to reach school. How many hours did it take to reach?

    _____

5. Mary-Ann read 4 pages in 8 minutes. How long does it take her to read 1 page?

    _____

    _____

6. Mommy's barbeque ham recipe requires her to bake for 4 hours. For how many minutes should she set the timer?

    _____

7. The pipe was left running for 420 seconds. How many minutes had it been running?

    _____

    _____

8. A fishing boat with 3 fishermen was in the water for 3 ½ hours. For how many minutes were they in the water?

_____

9. If it takes 8 hours to unload the bags with yam from the market truck, how many minutes does it take?

_____

_____

10. In a cross-country relay Shaun ran for 2 hours, Malik ran for 2 ½ hours and Omar ran for 1 ½ hours. How many minutes did they run in all?

_____

Evaluation: Let us see how you did.

Learning outcome	No!	Working on it	Yes!
Did you get all the answers?	☹	😐	🙂
Did you get most questions right?	☹	😐	🙂
Did you retry the question(s) you got wrong?	☹	😐	🙂
Were you able to correct your wrong answers?	☹	😐	🙂
If not, did you seek help from others and/or review the chapter?	☹	😐	🙂

Colour the face that shows how you are doing.

CHEETAH
Connect to Higher Education, Electronic Tools, Aplication and Help

What are the relationships between lines and angles?

Before we begin, let's see what you know.

Prior learning:

✓ Identify and describe a point, line segment, simple closed path, square corner.

Key vocabulary

Check the words you understand:

☐ angle
☐ intersecting
☐ line segment
☐ parallel
☐ perpendicular
☐ ray
☐ right angle
☐ turn

1. Match each word in column A with the correct definition in column B by drawing a line from one column to the next.

Column A	Column B
point	point where two-line segments meet at a right angle (90 °)
line segment	a location in space, usually represented by a dot ●
simple closed path	a straight path between two points A B A B
square corner	a line that starts and ends at the same point

If you know these, you should be able to learn the following.

🎯 8.1. Differentiate between concepts of point, space, curved/horizontal/ vertical/oblique lines or line segments.
8.2. Identify and name rays and associate them with the formation of angles.
8.3. Investigate the idea of a 'turn' and associate it with the formation of an angle.
8.4. Use capital/common letters to name angles/rays.
8.5. Recognize right angles when drawn or seen in the environment.
8.6. Use estimation to identify angles less than, greater than or equal to a right angle.
8.7. Identify angles from different perspective and orientations
8.8. Identify parallel, perpendicular and intersecting lines when drawn or seen in the environment.

8.1 Differentiate between concepts of point, space, curved/horizontal/ vertical/oblique lines or line segments.

Features of Geometry	Description
point	a location in space usually represented by a dot ●
space	area occupied by an object or shape; Area between two points
square corner or **perpendicular** lines	point where two **line segments** meet at a **right angle** (90 ^{0})
lines	a set of points moving away from a common point in opposite directions forever (has arrows on both ends) or horizontal line or **parallel** lines ‖ vertical line
line segment	a set of points connected by a straight path. A line is usually marked by 2 points. Example: line AB A B A B
curved line	a set of points placed at an angle behind each other that when viewed together form a bent line. Curved lines are not straight. An arc is an example of a curved line
oblique lines	a line that is not vertical or horizontal. Oblique lines are diagonal

8.2 Identify and name rays and associate them with the formation of angles.

A **ray** is a part of a line that starts at a point (end point). The side with the arrow goes on forever in one direction.

It is visually represented as ●————————▶

Let us practice 8.1 & 8.2

1. Which of these will form a square corner?
   a. A vertical line meets an oblique line.
   b. A horizontal line meets an oblique line.
   c. A horizontal line meets a vertical line.
   d. Two vertical lines meet.

2. Select the type of line that is used to form a circle.
   a. horizontal line
   b. vertical line
   c. oblique line
   d. curved line

3. Which of these ideas describe a line segment?
   a. a curved set of points
   b. a straight row of points
   c. a line labelled with one letter only
   d. a line made from two oblique lines

4. Which option defines the word *space*?
   a. crossing of 3 curved lines
   b. area enclosing a slope
   c. point where two lines cross
   d. area between 2 points

5. If a straight line passes through 3 points, what kind of line can it not be?
   a. curved line
   b. oblique line
   c. vertical line
   d. horizontal line

6. Which of these definitions is not true?
   a. line segment – part of a line with 2 end points
   b. space – a set of points
   c. point – a location in space
   d. shape – simple closed path

7. Look at this figure, ABCD. Which line is an oblique line?

   a. line AB
   b. line BC
   c. line DB
   d. line CD

8. A set of students labelled the diagrams below. Which pair of labels is incorrect?

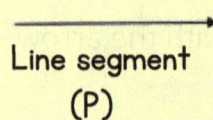

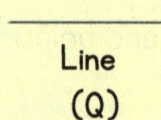

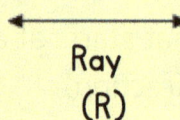

Line segment (P)	Line (Q)	Ray (R)	Point (S)

   a. P and S
   b. Q and R
   c. Q and S
   d. P and R

CHEETAH™
Connect to Higher Education, Electronic Tools, Aplication and Help

9. Which of the following types of line is not shown in the drawing?

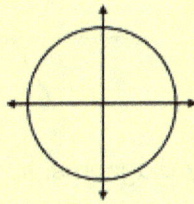

a. vertical line

b. horizontal line

c. curved line

d. oblique line

10. How many rays can be made from one line segment?

a. 4   b. 2   c. 8   d. 6

8.3 Identify and name rays and associate them with the formation of angles.

Rays form angles at the point from which they begin. In geometry, when two rays have a common starting point, an angle is formed between.

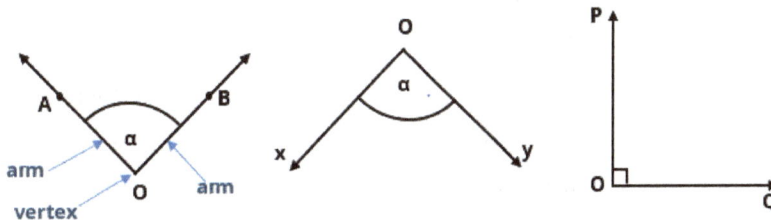

The vertex of the angle is the starting point of the rays, O (origin).

The angles are measured in degree (^{0}) between the two arms. The lengths of the rays do not affect the size of the angle between them.

Note angles 90^0 and 45^0.

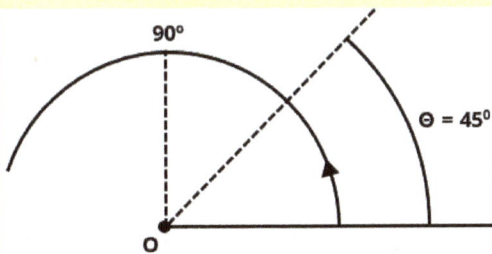

$\Theta = 45^0$

A protractor can be positioned and used to measure an angle as shown.

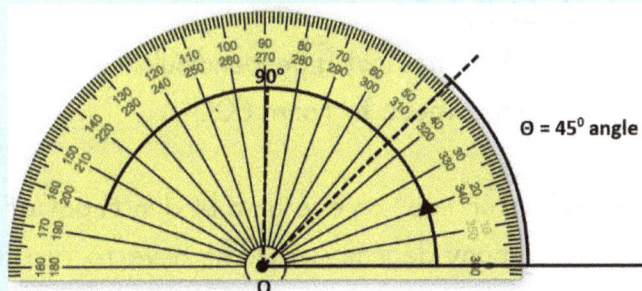

$\Theta = 45^0$ angle

Let us practice 8.3

1. How many rays do you observe in the drawing?

   a. 2   b. 4

   c. 6   d. 8

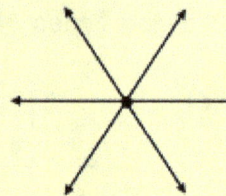

2. Practise drawing rays by completing these ray diagrams, label the points and name the angles formed.

   a. 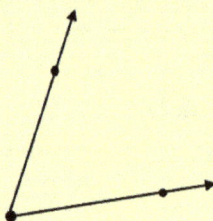   b. 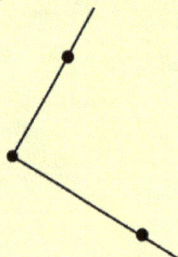   c. 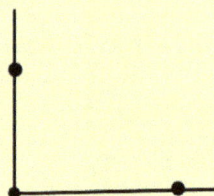   d.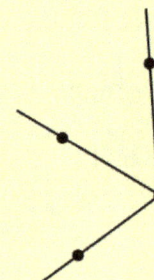

3. Complete the ray diagram, then label and name 8 angles using the common end-point.

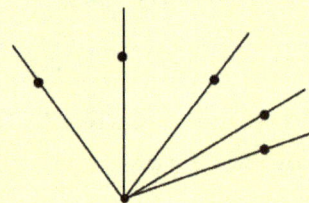

4. Which statement about rays is true?

   a. The number of rays that can leave a point can always be counted.

   b. As more rays leave a point, the angle between the rays gets smaller.

   c. As more rays leave a point the angles between the rays get larger.

   d. The angles are aways the same.

5. The diagram shows two rays as they leave point O, one ray vertical (OA) and one ray horizontal (OB). Another ray OC passes through the middle of angle AOB

   i. Draw the ray OC on the diagram.

   ii. What angle does this ray make with the ray line OB?

   a. 60°   b. 30°   c. 45°   d. 90°

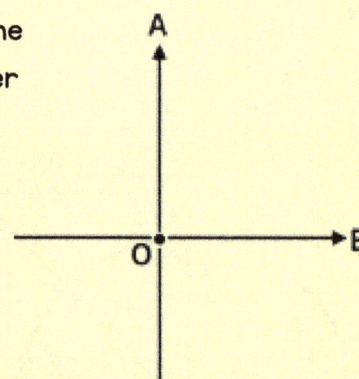

6. One ray is moving vertically, while another is moving horizontally. What is true about them?

   a. They meet at a perpendicular.   b. They will never meet.

   c. They meet to form 60°.   d. They meet to form a straight line.

7. A hunter kept an eye on a bird in front of him. The bird flew directly behind him. How many degrees would he have to turn to see the bird again?

   a. 0°   b. 90°   c. 180°   d. 270°

8.4 Investigate the idea of a 'turn' and associate it with the formation of an angle.

Draw a cross on the floor and stand at the centre point. While at the centre point there are 4 rays to look at: one in front, one behind you, one on the right and one on the left.

Look at the picture of the beach below. Stand in the centre O, facing the ray going to the 0^0, **turn** to another ray pointing to the car. Note each turn is 90^0 (a square corner). Turn to the sandcastle at 180^0 from 0^0.

One complete turn is $4 \times 90^0 = 360^0$. Note a complete turn is 360^0.

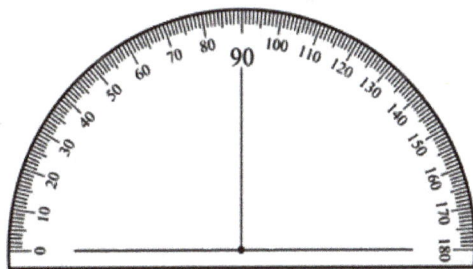

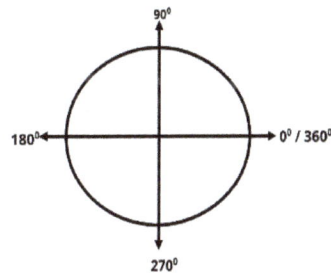

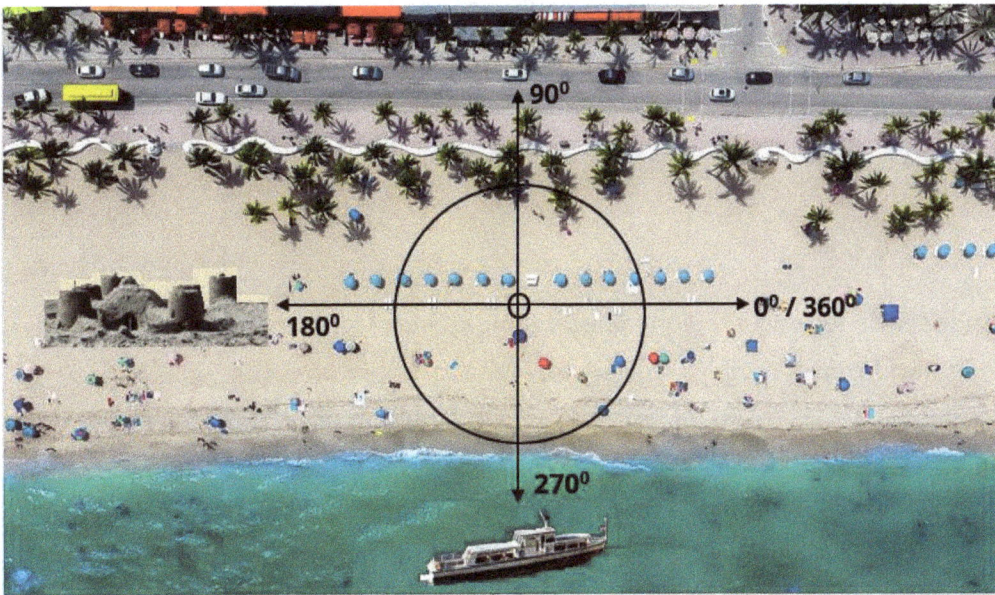

Let us practice 8.4

1. One complete circle is _____.
   - a. 180^0
   - b. 360^0
   - c. 160^0
   - d. 45^0

2. A quarter turn is _____.
   - a. 90^0
   - b. 45^0
   - c. 120^0
   - d. 60^0

3. If you move three-quarters of a circle, how many degrees have you moved?
   - a. 380^0
   - b. 270^0
   - c. 120^0
   - d. 56^0

4. The minute hand of the clock moves from numeral 12 to numeral 6. How many degrees has it moved?
    a. 240^0 b. 360^0
    c. 180^0 d. 85^0

5. If there are 12 numbers on the face of the clock that makes 360^0, how many degrees does each mark make?
    a. 60^0 b. 30^0 c. 36^0 d. 45^0

6. What is the angle between the hour and minute hands of the clock shown?

7. A driver reaches a main road and turns left at the T junction. What angle does the turn make?

8. If the angle between start and finish is a right angle on the diagram, what is angle b?

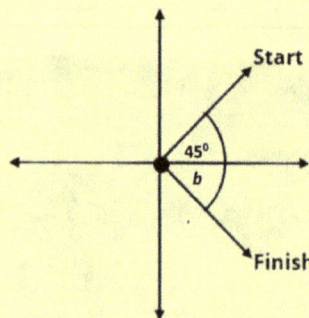

9. A dancer was looking straight ahead. She looked right then left. How many degrees did her head move in all?
    a. 60^0 b. 90^0 c. 180^0 d. 70^0

10. A wheel turned I complete revolution. How many degrees did it turn?
    a. 140^0 b. 120^0 c. 180^0 d. 360^0

8.5 Use capital/common letters to name angles/rays.

Look at the rays $\overrightarrow{OA}$ and $\overrightarrow{OB}$. They are joined at point O to make angle d. The common letter d is used to name the angle.

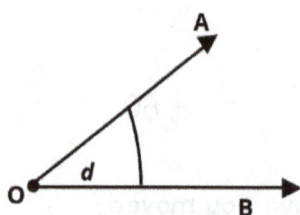

This angle named d can also be named using capital letters $A\hat{O}B$ and $B\hat{O}A$ shown in the diagram.

Note that angle $A\hat{O}B = d$ and $B\hat{O}A = d$,

so $A\hat{O}B = B\hat{O}A$

Practise drawing and naming rays using capital and common letters.

1. Label angles $\overrightarrow{AOC}$, $\overrightarrow{COD}$ and $\overrightarrow{DOE}$ using common letters.

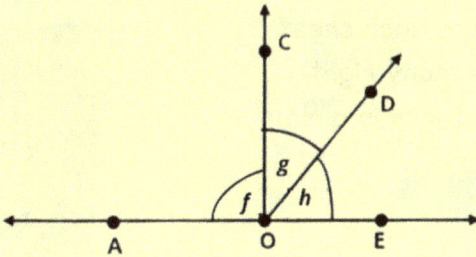

$\overrightarrow{AOC}$ = _____

$\overrightarrow{COD}$ = _____

$\overrightarrow{DOE}$ = _____

2. Complete the diagram by adding capital and common letters.

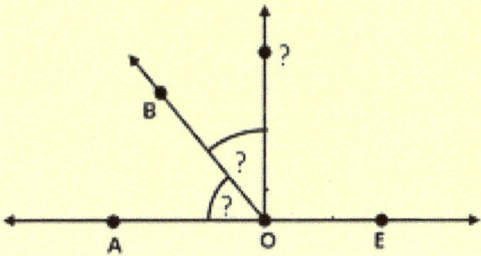

3. Name the angle labelled p and q using capital letters.

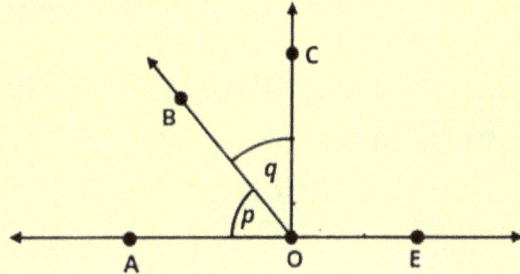

8.6 Recognize right angles when drawn or seen in the environment.

A right angle is formed when two lines meet at a 90⁰ angle. These two lines can be a horizontal and a vertical line or 2 oblique lines. They form a perpendicular or a square corner given as a small box in between the lines that meet.

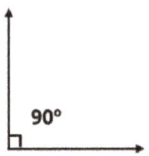

Square corner formed from a vertical and a horizontal line.

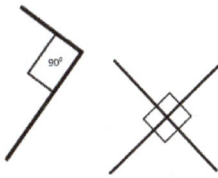

Square corner formed from oblique lines.

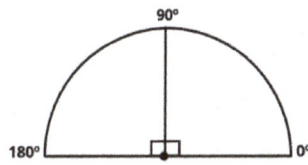

Square corner formed within a semi-circle from a vertical and a horizontal line.

1. For the table and chair given, draw square corners in the locations of the right angles.

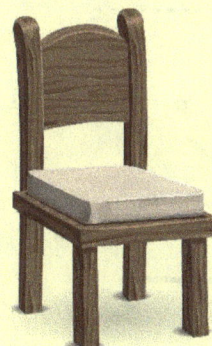

2. Given a box, circle all the right angles.

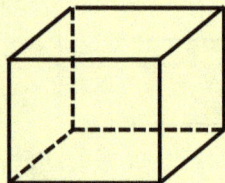

3. There are 9 sheets of paper. How many square corners does each sheet have? How many right angles are there in all?

4. Identify the types of angles seen on the shelf.

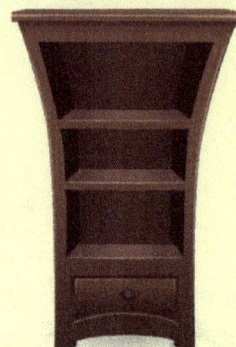

8.7 Use estimation to identify angles less than, greater than or equal to a right angle; Identify angles from different perspective and orientations.

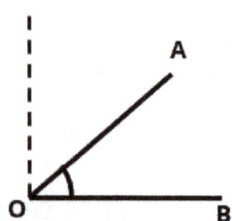

Acute angles – angles less than 90°

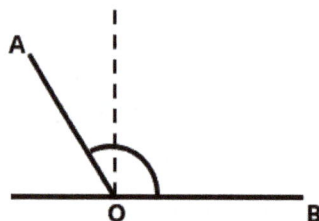

Obtuse angles – angles greater than 90° but less than 180°

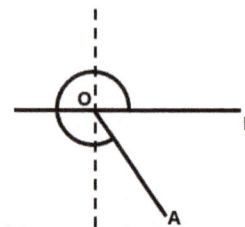

Reflex angles – angles greater than 180°, but less than 360°

Let us practice 8.7

1. Identify the drawings that show angles that are estimated to be right angles, acute angles, obtuse angles or reflex angles. Write your answer in the boxes for each drawing.

Note: θ - theta

a.

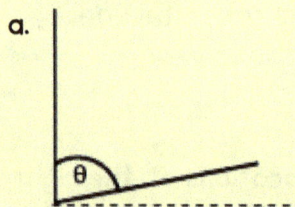

b.

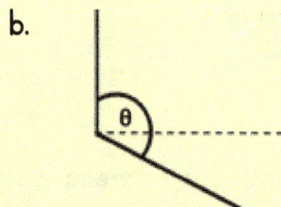

c.

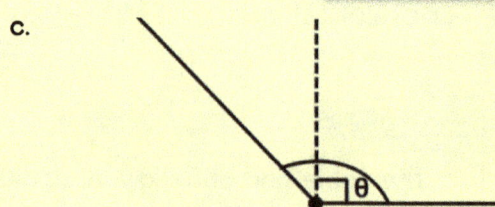

d.

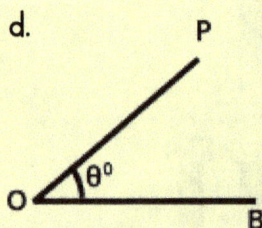

e.

f.

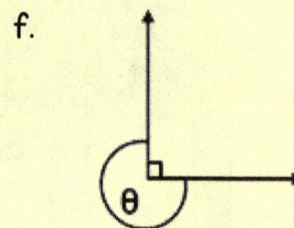

2. Identify, using letters, two acute angles, two obtuse angles, as well as two right angles in this drawing.

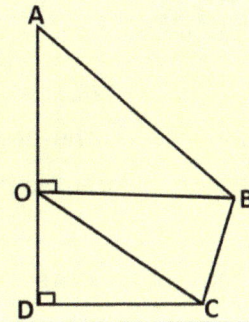

8.8 Identify parallel, perpendicular and intersecting lines when drawn or seen in the environment

Parallel lines: These lines are always the same distance apart, run in the same direction and never touch or cross each other.

Examples of parallel lines in our environment:

Perpendicular lines: These lines meet at a 90⁰ angle.

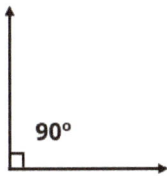

90°

Examples of perpendicular lines in our environment:

Intersecting lines: These segments meet or cross each other at one point. Four angles are formed when lines intersect. Angles formed on opposite sides of the X are equal.

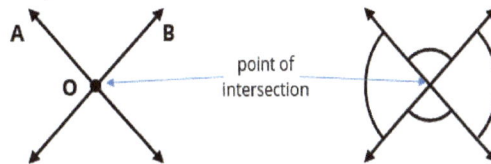

A   B

O   point of intersection

Examples of intersecting lines in our environment:

RAILWAY CROSSING

Examples of parallel, perpendicular and **intersecting** lines on a map of Kingston, Jamaica.

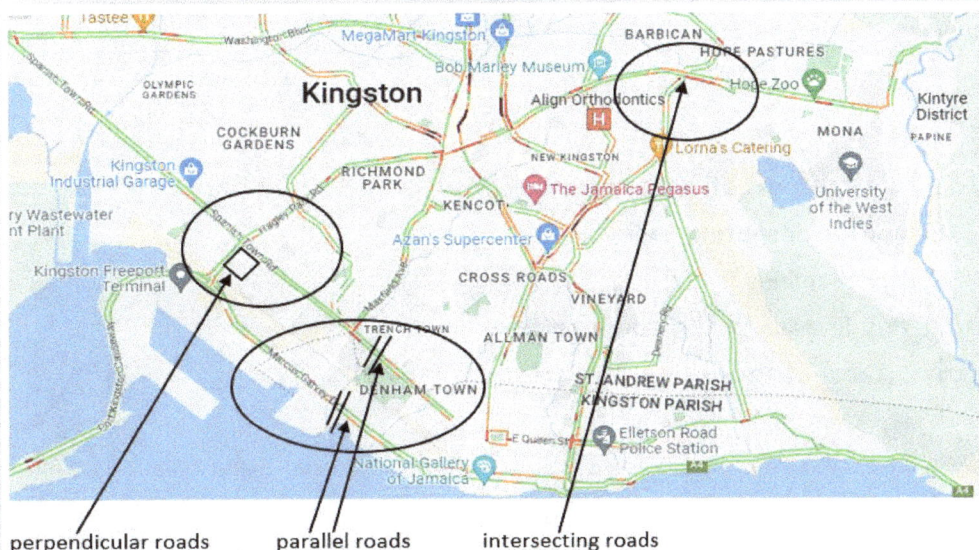

perpendicular roads   parallel roads   intersecting roads

Let us practice 8.8

1. What is always true about a right angle?
   a. The lines of the angle must be parallel.   b. The lines intersect.
   c. The lines do not come to a vertex.   d. The lines are vertical and horizontal.

2. Which statements are not true?

   i.   Intersecting lines do not form angles.
   ii.   Parallel lines form angles when they intersect.
   iii. A right angle is sometimes greater than 90°.
   iv.   All angles must have a vertex.

   a. i and ii   b. i, ii and iv   c. iv only   d. i, ii and iii

3. What must be true about angles?
   a. Both lines making an angle must be the same length.
   b. Angles do not need to be labelled.
   c. Angles are formed when rays leave a common starting point.
   d. Angles can be formed even when lines do not meet.

4. Which of these types of lines is used to draw a circle.
   a. oblique line   b. horizontal line   c. vertical line   d. curved line

5. If two lines do not intersect, then which is true?
   a. The lines cannot form an angle.   b. Both lines are rays.
   c. The lines will form an angle > 90°.   d. They form a very small angle.

6. Identify the oblique lines in this drawing.
   a. PQ and RS
   b. AC and BD
   c. MN and BD
   d. RS and AC

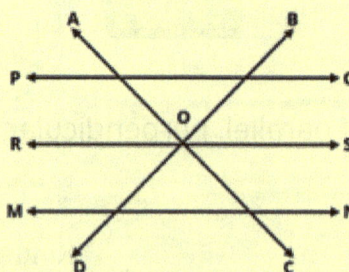

7. Identify the pair of lines given below.
   a. non-intersecting lines
   b. parallel lines
   c. ray lines with the same origin
   d. intersecting lines

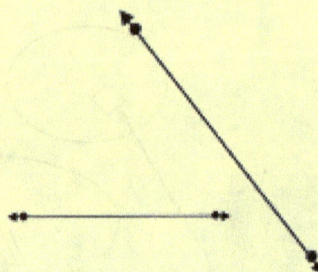

8. Look at this diagram. What type of line is not observed in the drawing?
   a. curved line
   b. oblique line
   c. horizontal line
   d. ray line

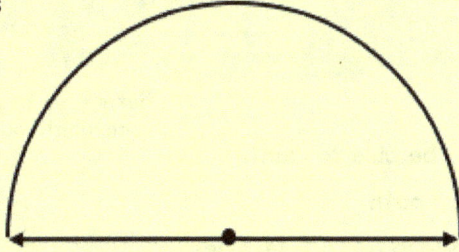

9. This is a racetrack at the stadium. Which type of line is not seen in the drawing?
   a. curved
   b. parallel
   c. oblique
   d. vertical

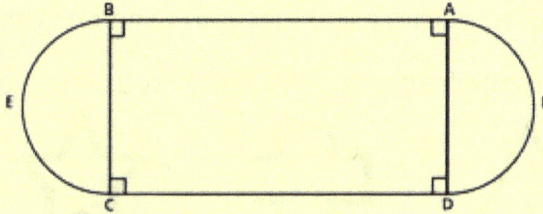

Evaluation: Let us see how you did.

Learning outcome	No!	Working on it	Yes!
Did you get all the answers?	☹	😐	🙂
Did you get most questions right?	☹	😐	🙂
Did you retry the question(s) you got wrong?	☹	😐	🙂
Were you able to correct your wrong answers?	☹	😐	🙂
If not, did you seek help from others and/or review the chapter?	☹	😐	🙂

Colour the face that shows how you are doing.

What are the similarities and differences among geometric shapes?

Prior learning:
Students should be able to identify:
- ✓ simple closed path
- ✓ polygons (having up to four sides)

Before we begin, let's see what you know.

A simple closed path is a line that starts and ends at the same point.

1. Which of the following does not show a simple closed path?

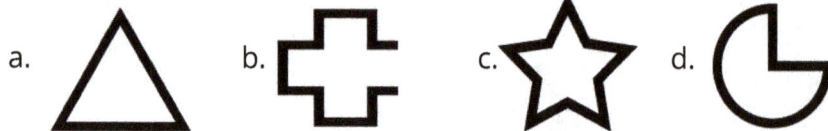

   a.   b.   c.   d.

A polygon with three sides is a triangle. A polygon with four sides is called a quadrilateral.

2. Use lines to match the names to the shapes of polygons with up to four sides.

   kite

   rectangle

   triangle

   square

If you know these, you should be able to learn the following.

Key vocabulary

Check the words you understand:

- ☐ regular polygons
- ☐ irregular polygons
- ☐ congruence
- ☐ congruent
- ☐ non-congruent
- ☐ polygons
- ☐ triangles
- ☐ squares
- ☐ rectangles
- ☐ quadrilaterals
- ☐ parallelogram
- ☐ rhombus
- ☐ kite
- ☐ trapezium
- ☐ hexagons
- ☐ pentagons
- ☐ septagons
- ☐ heptagons
- ☐ decagons
- ☐ octagons
- ☐ nonagons
- ☐ similar
- ☐ tangram

🎯 9.1 Identify congruent shapes and explain why they are congruent.
9.2 Differentiate between polygons and non-polygons.
9.3 Explore combinations of geometric shapes, especially triangles and quadrilaterals.
9.4 Identify and draw the following polygons: triangles, squares, rectangles and irregular quadrilaterals.
9.5 Draw pictures of a polygon to a reasonable degree of accuracy where the lengths of the sides or descriptions are given.

9.1 Identify congruent shapes and explain why they are congruent.

Congruent means that the figures (shapes) are the same size and shape. It is possible to place one shape over another and they match exactly. The shapes and size of congruent figures must be equal even if either figure is flipped, turned, or rotated.

Congruent **polygons** have the same number of sides, where the lengths and interior angles that correspond are equal.

In dealing with congruent figures, sides with the same number of strokes have equal lengths.

Example: Triangles

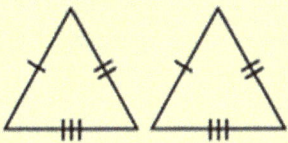

For these congruent figures, the sides with 1 stroke have the same length; the sides with 2 strokes have the same length; the sides with 3 strokes have the same length.

Example: Quadrilaterals

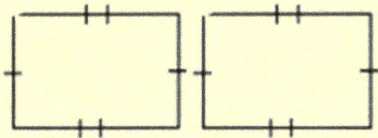

These figures fit exactly over each other.

Similarly, angles marked with the same number of curved lines are equal.

Example:

For these triangles, the angles with 1 curved line are equal; the angles with 2 curved lines are equal; the angles with 3 curved lines are equal.

Let us practice 9.1

1. Are these two angles congruent to each other? Explain your answer.

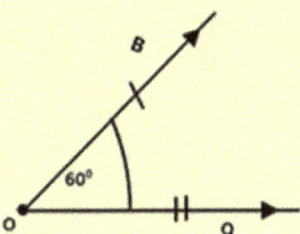

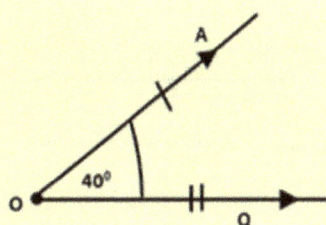

_____

2. Are these triangles congruent? Give a reason for your answer.

a.

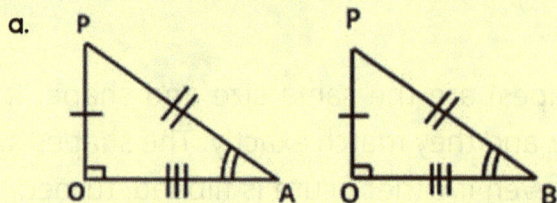

_____

_____

b.

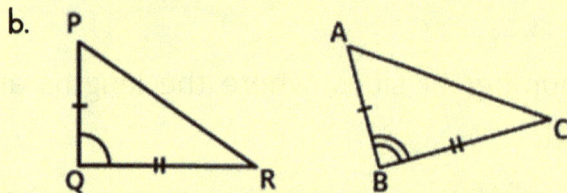

_____

_____

3. Are these figures congruent? Explain your answer.

a.

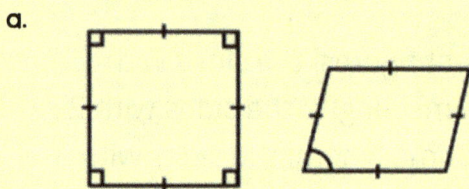

_____

_____

b.

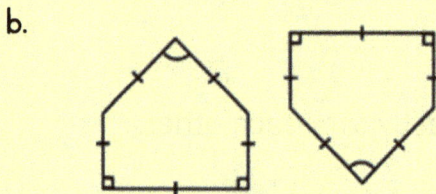

_____

_____

4. Are these triangles congruent? Circle yes or no. Explain your answer.

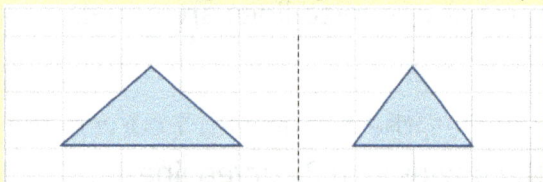

_____

_____

5. Draw two lines that can change this image into two different sets of congruent figures.

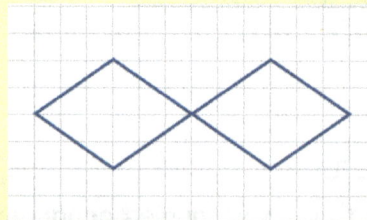

6. Are the pages of a book congruent? Explain your answer.

_____

_____

CHEETAH™
Connect to Higher Education, Electronic Tools, Aplication and Help

7. Look at these two houses in a housing scheme. Are the figures congruent? Explain your answer.

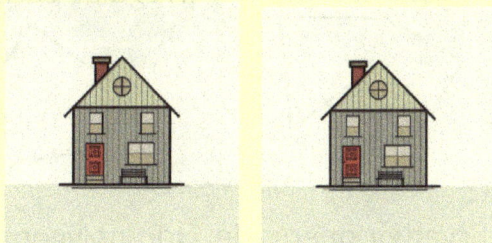

_____

_____

8. Are these figures congruent? Explain your answer.

A   B

_____

_____

9. Is the arrangement of the shapes congruent? Explain your answer.

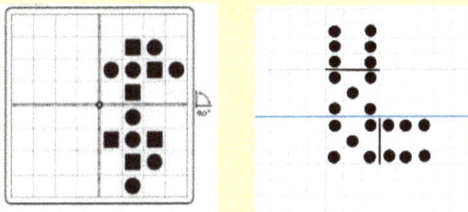

_____

_____

10. Are the 2 triangles congruent? Explain your answer.

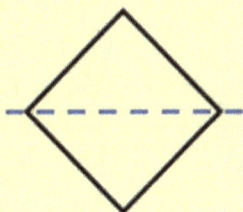

_____

_____

9.2 Differentiate between polygons and non-polygons.

A polygon is a closed figure with three or more vertices, angles and sides. All the angles and sides (line segments) in a regular polygon are equal. Polygons can be either regular or irregular. All the angles and the sides in an irregular polygon are not equal. For example, a square has equal sides and equal angles, so it is a regular polygon. However, a rectangle has only 2 of the 4 sides equal and all angles equal. So a rectangle is an irregular polygon.

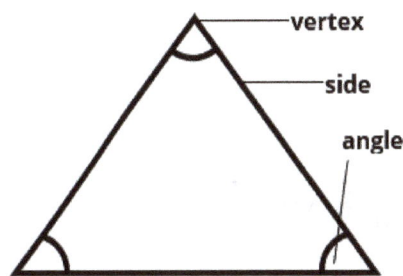

vertex
side
angle

Examples of irregular polygon are:

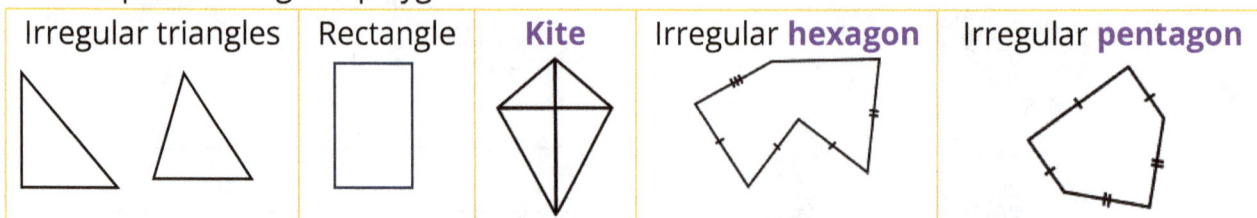

Irregular triangles	Rectangle	**Kite**	Irregular **hexagon**	Irregular **pentagon**

Non-polygons are shapes or figures with a curved or open side. Non-polygons are not polygons. Examples of non-polygon shapes include:

Let us practice 9.2

1. Complete the following table on properties of **regular polygons**.

Shape	Dimensions	Images
a. equilateral triangle	__sides, __ angles, __vertices	
b. square	__ sides, __ angles, __ vertices	
c.	5 sides, 5 angles, 5 vertices	
d.	__ sides, __ angles, __vertices	
e. heptagon	__ sides, __ angles, __ vertices	
f. octagon	__sides, __angles, __vertices	
g.	9 sides, 9 angles, 9 vertices	
h. decagon	10 sides, 10 angles, 10 vertices	

2. Which figures are polygons?

a. □　　　b. ○　　　c. 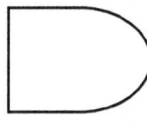　　　d.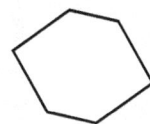

3. Look at this cube. How many regular 4-sided figures does it have?

Answer:_____

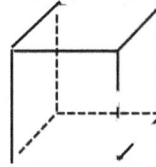

4. Use the letters to name as many regular and irregular 4-sided polygons as you can find.

Regular polygon:_____

_____

Irregular polygon:_____

_____

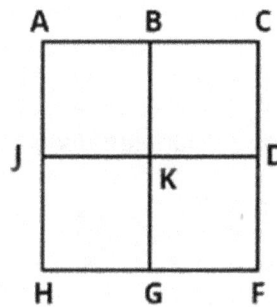

5. Which 2 shapes are used to make this figure?

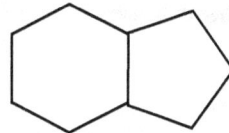

6. Look at the shape. Name the shapes listed using the letters at the vertices.

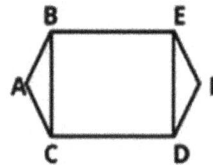

△　　　□　　　⬠　　　⬡

a. _____　b. _____　c. _____　d. _____

7. If you wanted to build a kite, how many pairs of congruent triangles would you need to paste together?

Number of pairs of congruent triangles = _____

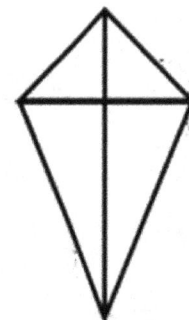

9.3 Explore combinations of geometric shapes, especially triangles and quadrilaterals.

To explore and identify combined geometrical shapes, you have to know and remember what they look like on their own. Remember, triangles are three-sided geometric shapes and quadrilaterals are four-sided geometric shapes. You can turn your page around to view the combined shape from a different angle or in a different way. You can also try numbering or colouring the shapes you see.

Here are some shapes you can practise on.

Let us practice 9.3

1. How many triangles do you see in the picture?

2. How many squares do you see?

3. What does a square need to become a hexagon? Draw a square then colour the triangles.

4. If a point is placed at the centre of a hexagon, how many triangles can be formed if a line is drawn from the point to each vertex? Draw the figure and count the triangles.

5. What shapes make up this figure?

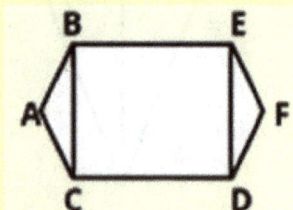

6. What shapes make up this figure?

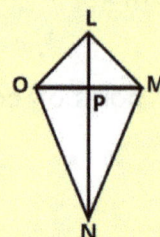

7. How many polygons can be seen in this figure?

8. Name and state how many triangles, rectangles, pentagons and hexagons you see in the figure?

triangles = _____   pentagons = _____

rectangles = _____   hexagons = _____

9.4 Identify and draw the following polygons: triangles, squares, rectangles and irregular quadrilaterals.

There are many different polygons with their many different number of sides. However, for grade 4 we will only draw the 3- and 4-sided polygons: triangles, **squares**, **rectangles** as well as regular and irregular **quadrilaterals**.

Let us practice 9.4

1. Name the following shapes, then tick whether they are regular or irregular.

	Shape	Name	Regular shape	Irregular shape
a.				
b.				
c.				
d.				
e.				
f.				
g.				
h.				

CHEETAH
Connect to Higher Education, Electronic Tools, Aplication and Help

9.5 Draw pictures of a polygon to a reasonable degree of accuracy where the lengths of the sides or descriptions are given.

Let us practice 9.5

Draw the following shapes using a ruler.

a. square with lengths 3 cm

b. right-angled triangle with sides 3 cm, 4 cm and 5 cm

c. rectangle with sides 2 cm and 5 cm

Evaluation: Let us see how you did.

Learning outcome	No!	Working on it	Yes!
Did you get all the answers?	☹	😐	🙂
Did you get most questions right?	☹	😐	🙂
Did you retry the question(s) you got wrong?	☹	😐	🙂
Were you able to correct your wrong answers?	☹	😐	🙂
If not, did you seek help from others and/or review the chapter?	☹	😐	🙂

Colour the face that shows how you are doing.

Prior learning:
✓ Collect and record data.

Before we begin, let's see what you know.

Key vocabulary

Check the words you understand:

☐ sample
☐ sample size
☐ population
☐ biased
☐ interview
☐ survey

1. Before school started in the morning and the parking lot was full of cars, a group of students were discussing what colour car was the most popular. Some students said red, others said white, black, yellow and blue. Each student had their favourite. To settle the difference of opinions, five students decided to come early to school and count and tally favourite colour cars seen for 30 minutes before school began. Their results were recorded in the table.

Colour	Tally	Total
white	‖‖ ‖‖ ‖‖	15
red	‖‖ |||	8
blue	|||	3
yellow	‖‖	5
black	‖‖ ‖‖	10

If you know these, you should be able to learn what comes next.

What these students have done is collect and record data.

Activity: Given the tally table of favourite fruits below, fill in the missing data.

Favourite fruits	Tally	Total
apples	‖‖ ||||	
plums		10
peas	||	
naseberry		6

🎯 10.1. Explain the idea of a 'sample'. 10.2. Explain the concept of 'population'.
10.3 Recommend a suitable sample size, based on a given scenario.
10.4 Determine whether a sample selected is appropriate based on the population.
10.5 Distinguish between a sample and a population as it relates to their sizes.

10.1 Explain the ideas of a population and a sample.

All the cars entering through the gate before school starts makes up the **population** of all the cars entering from the gate opens to when school starts.

The number of cars of different colours entering through the gate for 30 minutes before school starts would be a **sample** of all the cars entering through the gate before school starts.

Example 1: If there are 20 cookies in the jar and 2 are eaten, then the sample is 2 cookies and the population is 20 cookies.

Example 2: If there are 45 students in a class and 5 students are asked a question about their birthday, then the population is 45 students and the sample is 5 students.

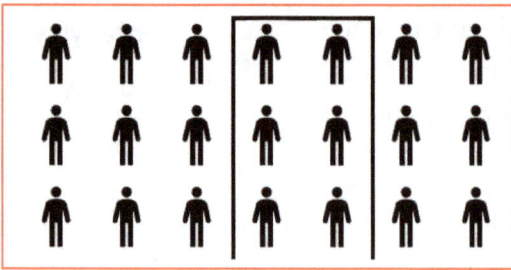

Example 3: If the population is in the red box, then the sample is in the black box.

The members of this sample were selected randomly. This means that no bias or preference was used to choose the members in the sample.

Example 4: A teacher wanted to know the colour marker that the class preferred. She did not ask everyone and take a tally. Instead, the teacher asked her four favourite students to choose the colour marker for the class.

This selection method was not random. The sample was selected with bias. To select four students at random, the teacher could have placed the names of all the students in a bag, shake it and blindly pick four names from the bag.

Let us practice 10.1

1. Annakay was told to sample the cookies and she ate all 5 on the tray. Was Annakay's action correct? Give a reason for your answer.

   _____

2. To help provide lunch at school, a teacher asked some students about their preference between chicken patties, cheese patties and beef patties. What is the population and sample from the information given.

   _____

3. There are 56 boys and 85 girls at school. What is the student population?

   _____

4. There are 8 classrooms and each classroom has 45 students. What is the school population?

   _____

5. Three hundred and four (304) grade six students did an exam. Eighty-five (85) students from the grade did not do the exam. What is the population of grade six students at the school?

_____

6. What is the population of flowers in the picture?

   Population = _____

   If only yellow flowers are in the sample, how many flowers make up the sample?

   _____

7. There were 140 people at a wedding. Of the attendees, 42 were hotel staff. The bride and the groom., and twenty-six (26) family members were also there. What was the guest population?

_____

10.2 Recommend a suitable sample size, based on a given scenario.

Sample size can be recommended if that sample will show a similar result as the population. A suitable minimum sample size that can be used to examine a population is 1 in every 10.

Example: If the first 100 cars entering a school were to be used to tell the favourite colour car, then a suitable sample size from the 100 is 10; that is, 1 car for every 10 cars in the population.

If there were 100 students, we asked 10 or more. If there are 800 students, we ask 80 students or more. The larger the sample size used, the closer the result would be to asking every student. The larger the sample, the more the sample would look like the population.

Let us practice 10.2

1. There are 40 students in a class. What could be the smallest sample to give an idea of how the class feels about eating chocolate cake?

   a. 10 students   b. 4 students   c. 6 students   d. 5 students

2. There are 60 jars containing jam on a table. How many jars of jam should be sampled to give a good idea if the jars have the volume of jam written on the label?

   a. 10 jars   b. 8 jars   c. 6 jars   d. 1 jar

3. A set of books has 20 pages in each book. A student wants to confirm if the books really have 20 pages in each book, so he counts the number of pages in 8 books as a sample for the population of books. Which statement best describes the number of books in the population?
   a. at least 80 books
   b. more than 80 books
   c. less than 80 books
   d. equal to 80 books

4. A student tested 3 pens in the box as a sample to make sure they worked. At least how many pens are likely to have been in the box? _____

5. A box of tissue has 50 rolls of tissues. Some students wanted to be sure that each had 100 squares. How many rolls should be counted as a sample?

   _____

6. Students measured the length of pencils to see if they were likely to be the correct length. If there were 70 pencils in the box, what would be the minimum sample size?

   _____

7. A case of 150 bottles of sweet drinks was marked 90 ml. How many bottles should be examined to give an idea of the volume in each bottle?

   _____

8. A man goes to buy ear plugs for resale. If he is buying 110 ear plugs, at least how many should be tested before the purchase? _____

9. Each large matchbox should have 140 sticks in the box. If the 13 sticks in a sample box were tested, by how many sticks was the box short? _____

10. The box of sweets has 50 sweets on the label. A student wanted to make sure most of them had 50 sweets. If there are 70 boxes, how many boxes should be counted?

    _____

10.3 Determine whether a sample selected is appropriate based on the population.

Recall that a good sample is 1 in every 10.

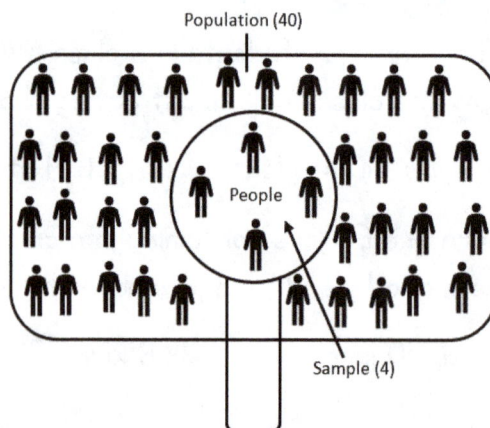
Population (40) — People — Sample (4)

Let us practice 10.3

1. Given the population size or the minimum appropriate sample size, complete the table by adding the missing data.

Population	Sample
80	
	6
140	_____
	12
320	_____

2. Given the population and sample amount, place a tick in the column to indicate whether the sample amount is good or not good.

	Population	Sample amount taken	Good sample	Not good sample
i.	200	16		
ii.	150	12		
iii.	90	10		
iv.	120	12		
v.	240	23		

10.4 Distinguish between a sample and a population as it relates to their sizes.

A sample is a small part (or fraction of a group), while the population is the whole. A sample gives an idea of what the whole group is like. The sample can tell what the population thinks and feels about some issue or idea.

Let us practice 10.4

1. The total population is 75. The number of women in the population is 35. How many men would you use in a good sample to answer the question, 'Do you often skip breakfast?'

   _____

2. In a population of 85 people, 45 are men. What would be a good sample size among the women to ask the question, 'Do you usually have soup on a Saturday'? _____

3. There are 80 peas. We would like to plant them, but we do not know if they will grow. How many would you select to test out the planting? _____

4. There are 150 men at church and a sample is to be chosen at random for the choir. Twelve (12) were chosen. Is this an appropriate sample?

   _____

5. There were 50 problems on the paper and the students were asked to do a suitable sample. The students had 10 problems. Was this a suitable sample?

   _____

6. A teacher wanted to know the students' favourite subject. Of the 30 students in the class, what is the least number of students that the teacher should choose as a sample to ask the question? _____

7. The tuck shop wanted to know which type of patties the students liked best. There were 300 students in the school. How many students should be sampled to give an idea of the type of patties to buy for resale? _____

8. The teacher wanted to form a football sports club but did not know how many students in school liked to play football. If there are at least 80 eligible boys, how many boys should be asked to get an idea of the popularity of football?

   _____

9. The teacher wanted to form a netball club but did not know how many students in school liked to play netball. If there are at least 120 eligible girls, how many girls should be asked to get to get an idea of the number of girls who are interested in joining a netball club? _____

10. At a road checkpoint the police were asked to stop a sample of the cars passing the point. If 400 cars pass the point, at least how many were checked?
   a. 34   b. 20   c. 40   d. 24

Evaluation: Let us see how you did.

Learning outcome	No!	Working on it	Yes!
Did you get all the answers?	☹	😐	🙂
Did you get most questions right?	☹	😐	🙂
Did you retry the question(s) you got wrong?	☹	😐	🙂
Were you able to correct your wrong answers?	☹	😐	🙂
If not, did you seek help from others and/or review the chapter?	☹	😐	🙂

Colour the face that shows how you are doing.

II.I. Collect numeric data based on interviews and observations.
II.2. Classify and sort data.

Data can be collected using interviews, observations, questionnaire (surveys) or even calculations. This data can be grouped or organised in tables, displayed in charts and graphs or listed in sentences.

11.1 Collect numeric data based on interviews and observation

We can use a table to show data collected. For example, when we collect data on the most popular coloured car entering a school within a 30-minute period, we may use observations to get the data and tables to show the data in an organized way.

When collecting data, honesty is very important. We should never falsify or change the data to look like what we think or what we prefer. Here are some of the ways that data can be collected.

Key vocabulary

Check the words you understand:

☐ tally marks
☐ data
☐ bar graph
☐ transportation

interview

survey questionnaire

research

audio and video recording

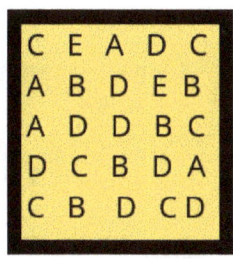
focus group discussion

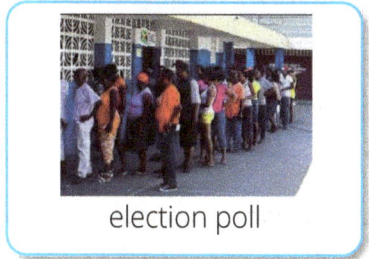
election poll

Part A: Collecting data

Example 1: Observe and collect data in a table.

How to create a tally table from data: Observe the Alphabet picture below. Use the picture to collect the data by sorting and tallying the letters with **tally marks** to complete the tally table.

```
C E A D C
A B D E B
A D D B C
D C B D A
C B D C D
```

Letter	Tally	Number
A	IIII	4
B	HHT	5
C	HHT I	6
D	HHT III	8
E	II	2

Tally table showing observations made from Alphabet picture.

Part B: Data organized in a table

Example 2: Interview questionnaire showing the **transportation** methods of 40 students.

Scenario - Forty (40) students were interviewed using a questionnaire. Each student was asked to record his/her method of transportation used to get to school. The table is a sample questionnaire completed to collect the data from each student.

Questionnaire					
Question asked: Which method of transport did you use to get to school?					
Motorcycle	Walk	Private car	Taxi	Bicycle	Bus
				✓	

This table shows how the data from the questionnaires are put together for all 40 students. Each stroke represents the answer given by each student who completed the questionnaire.

Question asked: Which method of transport did you use to get to school?		
Method	Tally	Total
walk	⣿⣿ IIII	14
private car	⣿ I	6
taxi	III	3
bicycle	⣿	5
bus	⣿ ⣿	10
motorcycle	II	2
Total		40

Let us practice 11.1

1. Do a survey to find your class's favourite type of cookies. The choices are strawberry, chocolate chip, vanilla, peanut, oatmeal and raisin cookies. Ask each student was asked to select one (1) favourite type of cookie.

Use this table to collect and tally the data on the types of cookies with the number of students who choose that type of cookie.

Type of cookies	Tally (Number of students)	Total
strawberry		
chocolate chip		
vanilla		
peanut		
oatmeal		
raisin		
Total =		_____ students

2. A class of students was planning to buy a pet. Each student was asked to recommend a pet. The pet with the most votes will be bought. The teacher recorded the data in a table. Note that each student could only vote once and every student choose a pet.

Complete this table to show what the data could look like.

Type of pet	Tally	Number
dog		
cat		
hamster		
bird		
fish		
	Total =	

3. Students at a school were asked to tick the type of cooked chicken they preferred. The table shows a tally for favourite types of chicken lunch at school.

Chicken lunch	Tally	Number
fried chicken and rice		14
stew chicken and rice		17
curried chicken and rice		22
fried chicken and chips		30
chicken soup		18
	Total =	102

4. Students went on a field trip to look at the different shaped leaves seen on plants in the school yard. Record the number of plants in your school yard with these shapes of leaves.

Different shaped leaves	Tally	Number
	Total =	

5. A bag had 20 jellybeans of different colours red, orange, yellow, blue and green which were not in equal amounts. Complete a tally and number table for an example of what this data may look like.

1.		
2.		
3.		
4.		
5.		
	Total =	

6. Do a summary to capture data on parishes in which parents of students in your class were born. Complete the table to show the data from 20 parents.

Parishes		
1.		
2.		
3.		
4.		
5.		
6.		
7.		
8.		
9.		
10.		
11.		
12.		
13		
14.		
	Total =	

11.2 Classify and sort data

Data can be classified by grouping like data together. For example, we can count all the letter C to group or classify them. When grouping Cs we do not count As because As are not Cs.

The letters in this alphabet picture are classified and sorted to give 4 letter A, 5 letter B, 6 letter C, 8 letter D and 2 letter E.

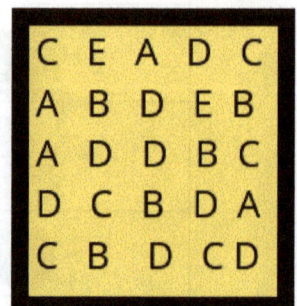

```
C E A D C
A B D E B
A D D B C
D C B D A
C B D C D
```

Letter	Number
A	4
B	5
C	6
D	8
E	2

This can be displayed as shown in the table, line graph and **bar chart** below.

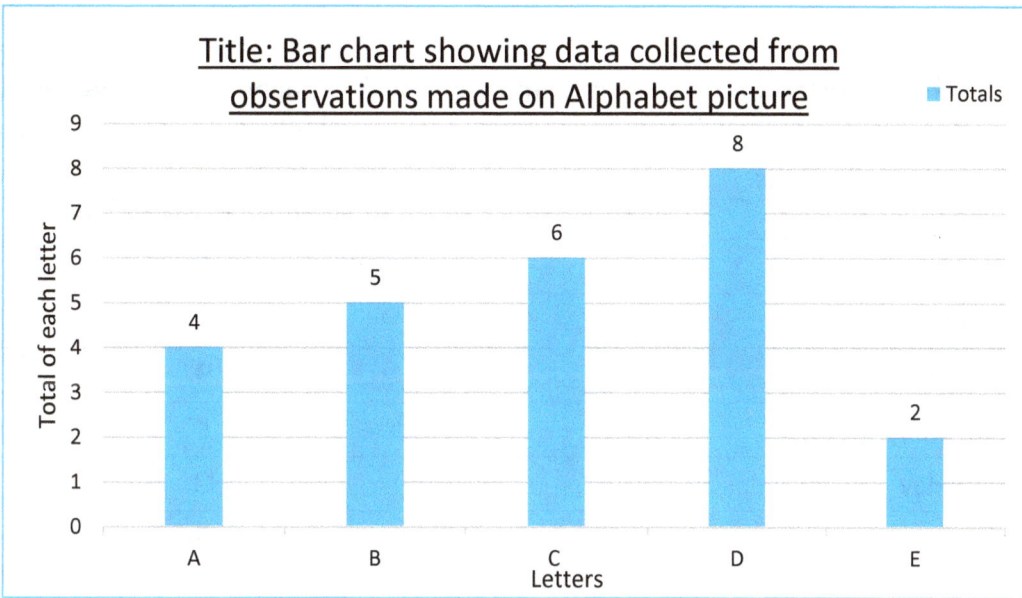

Title: Bar chart showing data collected from observations made on Alphabet picture

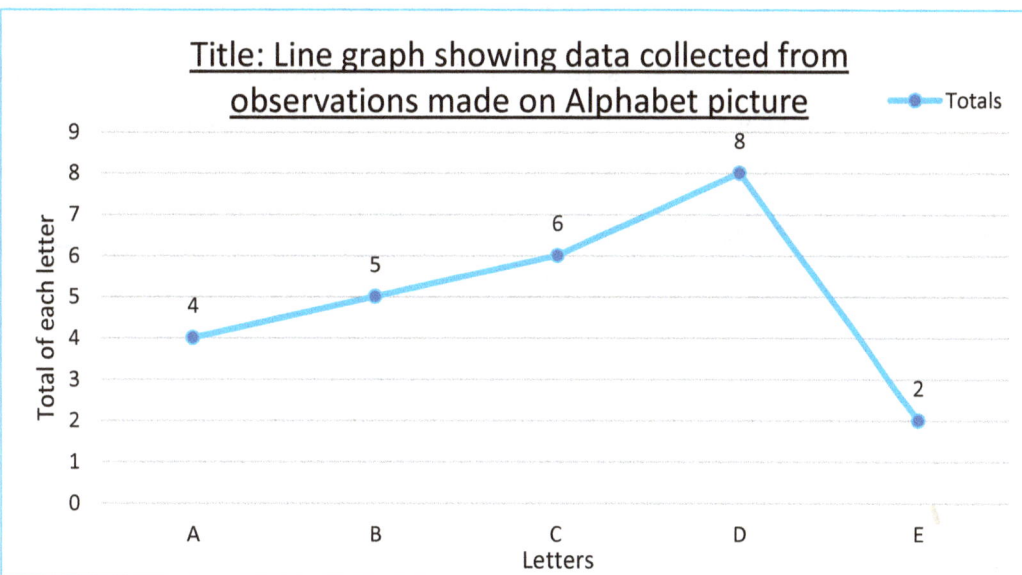

Title: Line graph showing data collected from observations made on Alphabet picture

Drawing inference based on the data

An inference is what we say about what we observe. It is the conclusion we get based on our reasoning or evidence. We draw inferences from our observations on the Alphabet picture shown as a table, a line graph or a bar chart by asking the questions:

1. What do we know?
2. What did we learn?
3. What inference can we make?

Inferences made:

a. Letter D was the most common.

b. Letter E was least common.

c. Letter D was at least 2 times more common than A and E.

d. Both vowels appear the same number of times.

e. The number of times A, B, C and D appear are at least 2 times that of letter E.

Let us practice 11.2

1. When the house phone bill came, mom was worried the phone calls made were not all ours so she decided to check the number of calls for the last week of the month. Complete the tally table.

Days	Tally	Number of calls per day
Sunday		6
Monday		8
Tuesday		9
Wednesday		11
Thursday		13
Friday		9
Saturday		18

2. A survey was done on the subjects to see which students preferred. Each student was allowed to vote once for each subject. The results of the survey were drawn on a line graph.

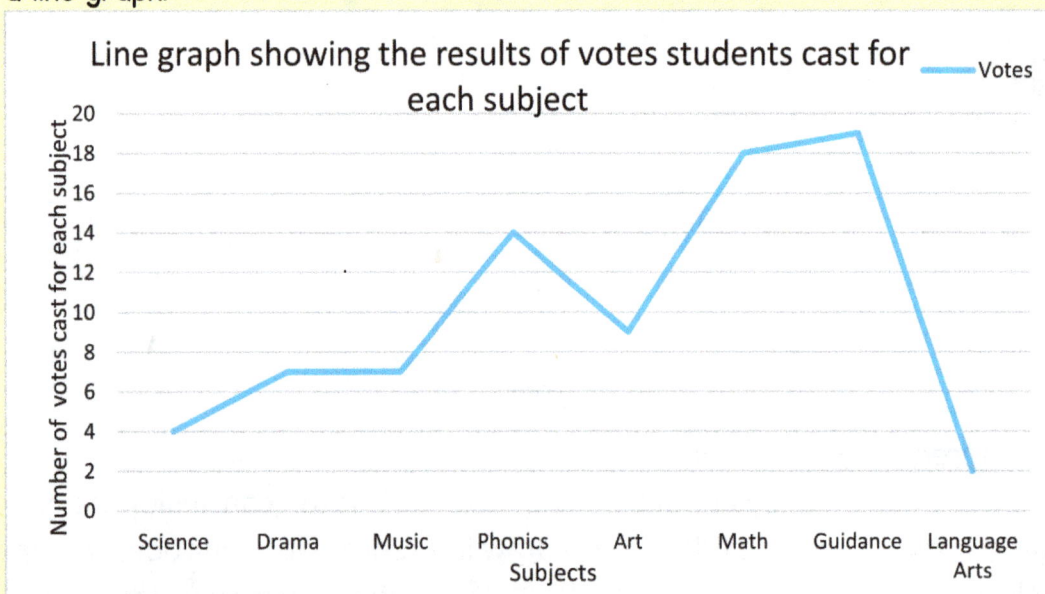

i. Use this line graph to draw a table with the subject name, tally and total number of votes.

ii. Draw a bar chart to show the data.

iii. Make inferences regarding the students' subject preferences using the presented data.

iv. Describe the information shown by the graph and bar chart.

3. The students did a survey on the size shoes that the classmates wore and collected the data given.

   i. Complete the table.

Shoe size	Tally	Number of shoes
6	IIII	
7	HHH I	
8	HHH III	
9	HHH II	
10	HHH	
11	I	
	Total =	

ii. Use a ruler to draw a bar chart for the data above. Remember to write in your title.

Bar Chart on _____

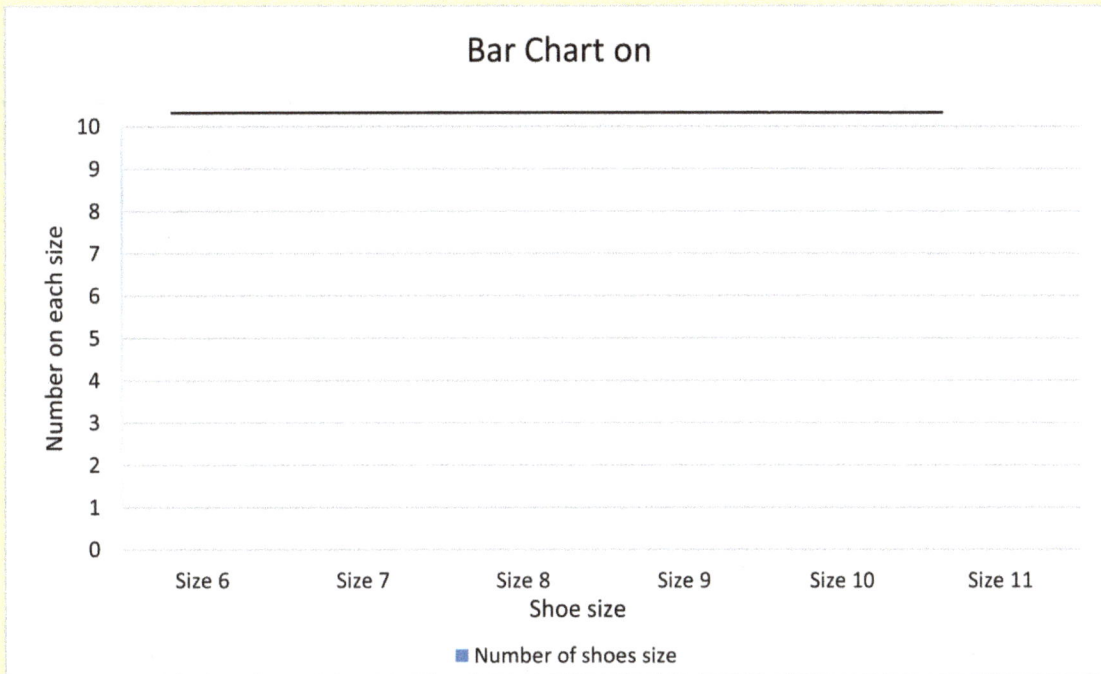

Number on each size

10, 9, 8, 7, 6, 5, 4, 3, 2, 1, 0

Size 6   Size 7   Size 8   Size 9   Size 10   Size 11

Shoe size

■ Number of shoes size

4. This table shows the favourite colours of 40 students. Complete the table and use it to draw a bar chart for the data.

Favourite colours	Number of students
red	7
yellow	8
blue	10
purple	9
green	6
Total =	

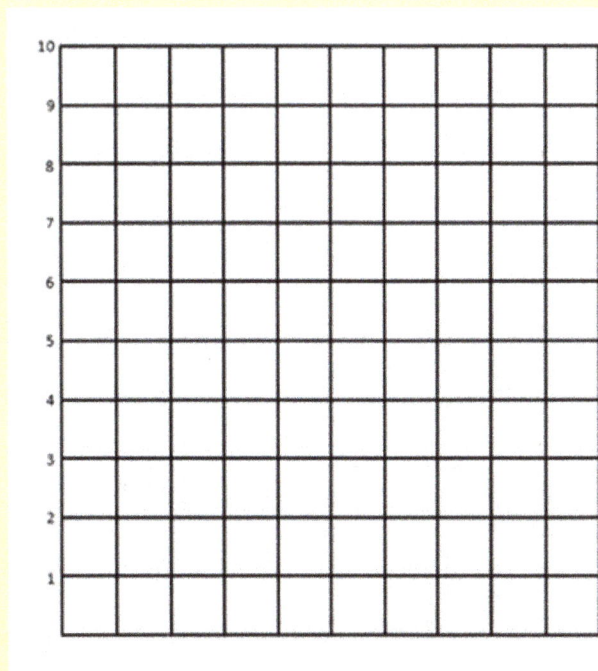

5. Complete the table showing the number of different types of vegetable plants bought at a farmers' market. Use the table to draw a bar chart for the data.

Types of vegetables	Number of vegetable plants
cabbage	2
lettuce	_____
spinach	6
broccoli	8
callaloo	10
Total	30

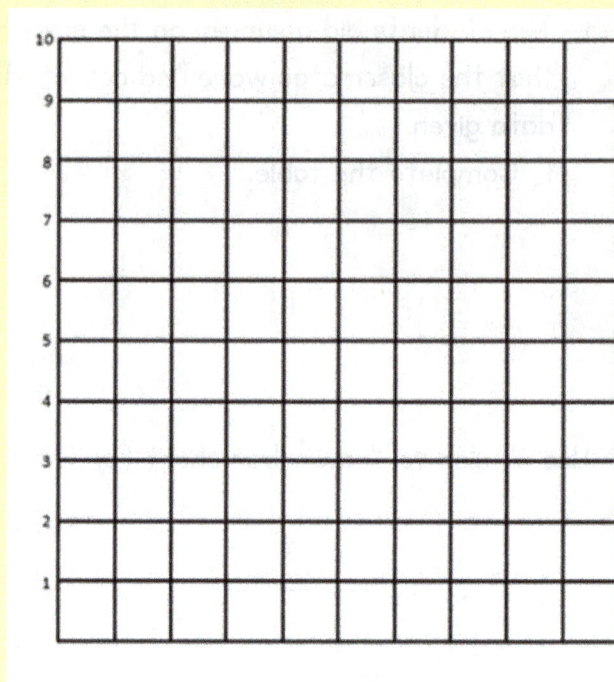

6. Complete the table showing the types of books borrowed from a library where 25 books were borrowed in all. Use the table to draw a bar chart for the data.

Types of books	Number of library books borrowed
comic books	7
magazine	_____
school textbooks	2
storybooks	10
journals	1
Total = _____ books	

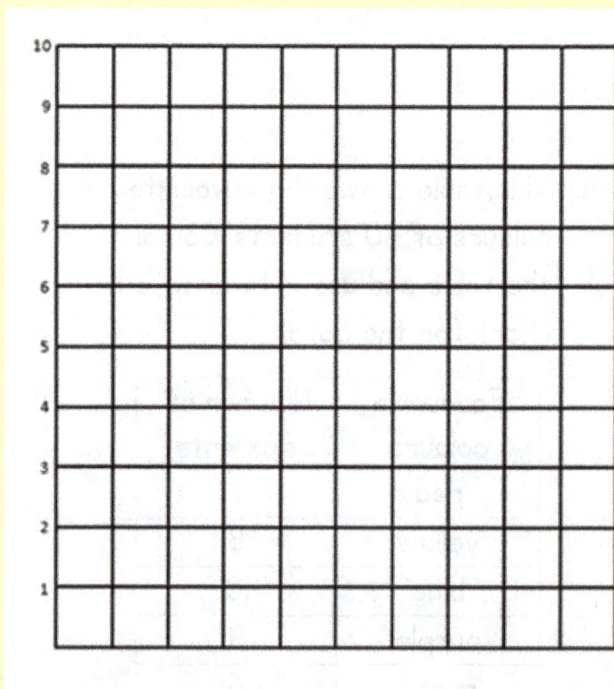

7. Complete the table showing the types of movies preferred by a youth group in a survey. Use the table below to draw a bar chart for the data.

Table title: _____

Types of movies	Number of youths preferring each type of movie
comedy	10
drama	5
romance	8
horror	3
adventure	8
Total =	_____

Title: _____

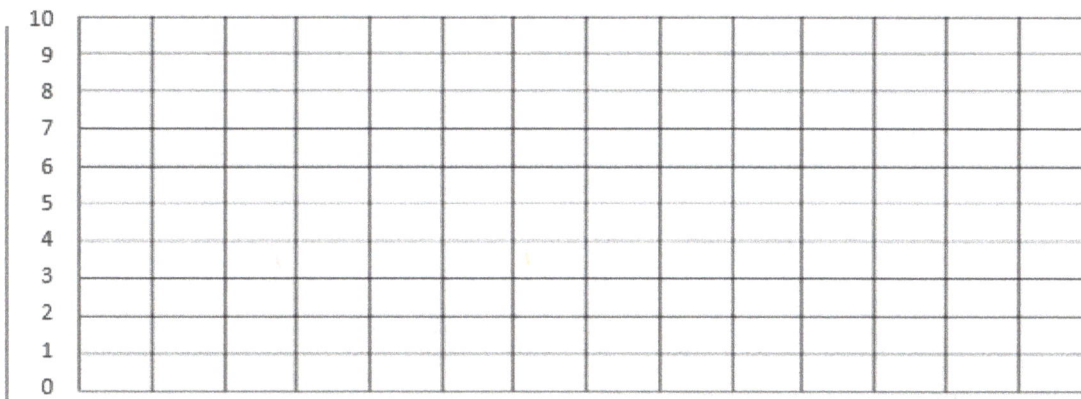

_____

8. A student wanted to know where to buy at least 5 types of sweets. Where would be the best place(s) to shop (Sh1 –Sh5)?

_____

_____

_____

Line graph showing 5 shops and the number of types of sweets each sold

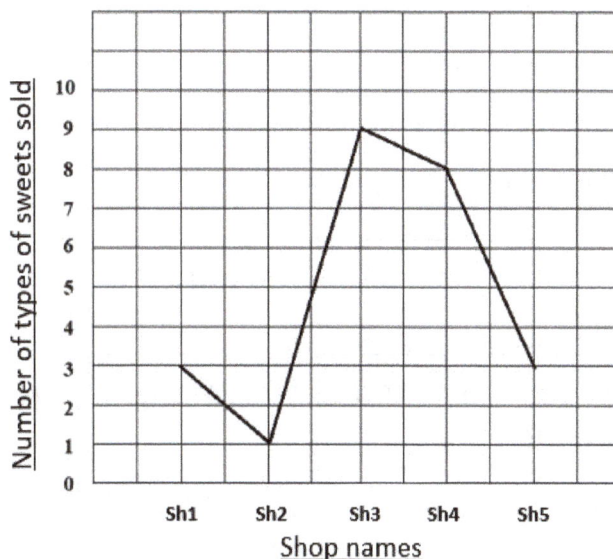

9. Twenty-eight (28) students out of the school's grade 6 population were asked to record the hours they spent on social media daily on a questionnaire.

   i. Complete the tally table for the results.

Time spent daily on social media (hrs.)	Tally of students	Number of students
1		10
2		7
3		6
4		3
5		2

   ii. Draw a bar graph to display the data in the table.

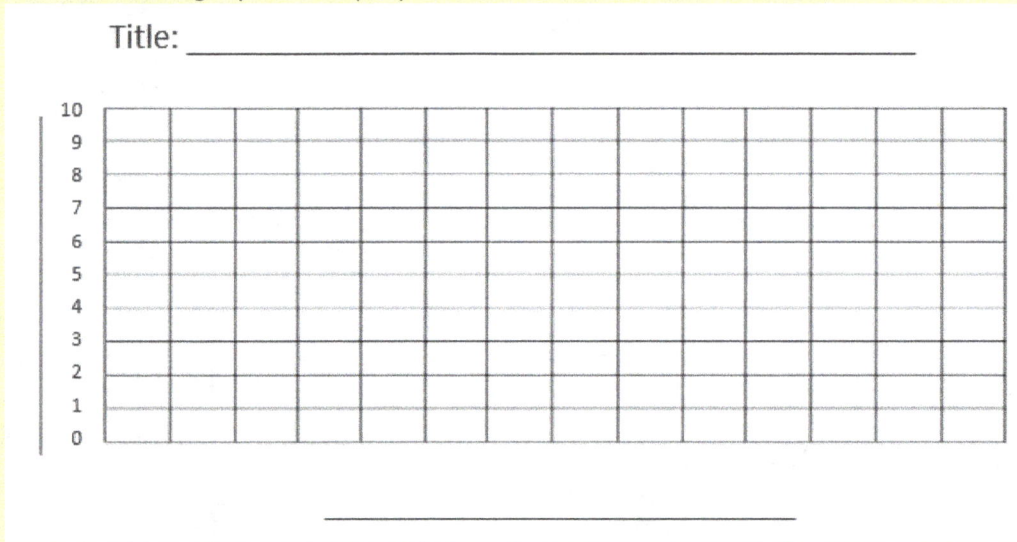

Title: _____

_____

   iii. What conclusions can be drawn?

_____

Evaluation: Let us see how you did.

Learning outcome	No!	Working on it	Yes!
Did you get all the answers?	☹	😐	☺
Did you get most questions right?	☹	😐	☺
Did you retry the question(s) you got wrong?	☹	😐	☺
Were you able to correct your wrong answers?	☹	😐	☺
If not, did you seek help from others and/or review the chapter?	☹	😐	☺

Colour the face that shows how you are doing.

TERM 2

$$52 = 50 + 2$$

$$52 \times 5 = (50 + 2) \times 5$$
$$= 50 \times 5 + 2 \times 5$$
$$= 250 + 10$$
$$= 260$$

How can I apply multiplication and division to larger numbers?

Before we begin, let's see what you know.

Prior learning:

✓ Recall multiplication facts up to 4 times.
✓ Identify pairs of related multiplication facts.
✓ Multiply a 3-digit number by a 1-digit number.
✓ Use the commutative property.
✓ Arrange sets of items using arrays.
✓ Compute with whole numbers (up to 3 digits).
✓ Estimate and check the reasonableness of answers.

Key vocabulary

Check the words you understand:

☐ breath
☐ breathing
☐ breathe
☐ estimation
☐ health
☐ million
☐ multiplier
☐ physical exercise
☐ product

4. Recall multiplication facts up to 4 times by completing the following multiplication table.

Times	1	2	3	4	5	6	7	8	9	10	11	12
2		4							18			
3				12							33	
4							28					

2. Identify pairs of related multiplication facts by completing the following table.

a.	2 x 4 =		4 x 2 =	
b.	3 x 4 =		4 X 3 =	

c. Multiply a 3-digit number by a 1-digit number.

a.	b.	c.	d.
768 x 2	576 x 3	896 x 4	976 x 5

d. Using the commutative property, a + b = b + a. Also, a x b = b x a or ab = ba.

Using commutative property, select the true statements.

a. a x b = b x a _____
b. a + b = b + a _____
c. 2a + 2b = 2ab _____
d. 3ab = 3a + 3b _____
e. a + c + b = a + b + c _____

Note that the same is not true for subtraction or division. That is a − b ≠ b − a. And $\frac{a}{b} \neq \frac{b}{a}$. Only when both a is 1 and b is 1 is the division true. So, $\frac{4}{2} \neq \frac{2}{4}$ and $\frac{6}{2} \neq \frac{2}{6}$

So, the commutative law works for addition and multiplication, but does not work for subtraction and division.

5. Arrange sets of items using arrays.
   An array is a set of objects that is set out in rows and columns. The table shows coloured blocks arranged in an array.

   For questions (a) and (c) use the following images.

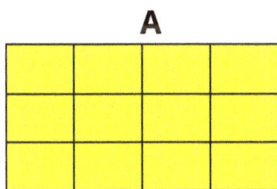

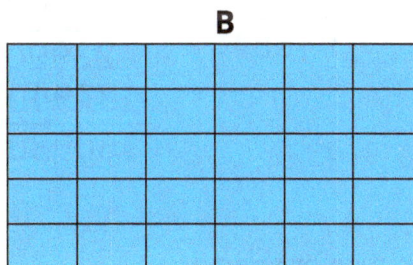

   a. How many small boxes do you see in A and B? Can you tell without counting all the boxes?

   _____

   b. Anthony planted 5 rows of pepper plants with 6 plants in each row. How many pepper plants did he plant.

   _____

   c. There are 6 benches in the class and each bench has 4 students. How many students are in the class?

   _____

6. How many soldiers are on parade?

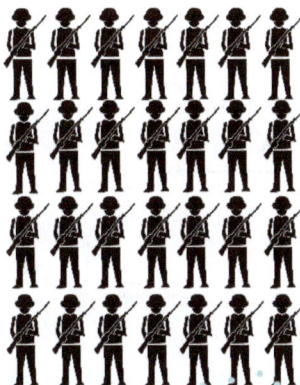

   If you know these, you should be able to learn what comes next.

12.1 Multiply numbers of up to four digits by any one- or two-digit number (including money).
12.2 Reinforce the mental multiplication of two-digit numbers by one-digit numbers.
12.3 Multiply a number by multiples of ten.
12.4 Identify and correct wrong answers in problems involving multiplication.
12.5 Discover, memorize and recall all multiplication facts up to at least 12 x 12 = 144.
12.6 Differentiate between the use of addition and multiplication, subtraction and division in problem situations involving whole numbers.
12.7 Select data relevant to a problem when finding its solution.
12.8 Identify and solve two-step problems.
12.9 Write mathematical sentences for a two-step problem.
12.10 Estimate and check answers to computations/problems.

12.1 Multiply numbers of up to four digits by any one- or two-digit number (including money).

Examples:

$$\begin{array}{r} 2367 \\ \times\ \ \ 4 \\ \hline 9468 \end{array}$$

$$\begin{array}{r} 4598 \\ \times\ \ 35 \\ \hline 22990 \\ 137940 \\ \hline 160930 \end{array}$$

Let us practice 12.1

1. Complete these multiplication problems.

a. $\begin{array}{r}6924\\ \times\ \ \ \ 4\\ \hline\end{array}$	b. $\begin{array}{r}7596\\ \times\ \ \ \ 5\\ \hline\end{array}$	c. $\begin{array}{r}9528\\ \times\ \ \ \ 6\\ \hline\end{array}$
d. $\begin{array}{r}3841\\ \times\ \ \ \ 7\\ \hline\end{array}$	e. $\begin{array}{r}2659\\ \times\ \ \ \ 8\\ \hline\end{array}$	f. $\begin{array}{r}5714\\ \times\ \ \ \ 9\\ \hline\end{array}$

2. Write the correct answers.

a. $\begin{array}{r}6958\\ \times\ \ \ 89\\ \hline\end{array}$	b. $\begin{array}{r}4569\\ \times\ \ \ 56\\ \hline\end{array}$	c. $\begin{array}{r}1374\\ \times\ \ \ 95\\ \hline\end{array}$
d. $\begin{array}{r}7986\\ \times\ \ \ 34\\ \hline\end{array}$	e. $\begin{array}{r}2037\\ \times\ \ \ 61\\ \hline\end{array}$	f. $\begin{array}{r}7060\\ \times\ \ \ 50\\ \hline\end{array}$

12.2 Reinforce the mental multiplication of two-digit numbers by one-digit numbers.

Memorize the following examples. Use this method to mentally solve the practice questions below.

Distributive property (a + b) c = a x c + b x c		
Note: 65 = 60 + 5	Note: 52 = 50 + 2	Note: 39 = 30 + 9
65 x 8 = (60 + 5) x 8 = 60 x 8 + 5 x 8 = 480 + 40 = 520	52 x 5 = (50 + 2) x 5 = 50 x 5 + 2 x 5 = 250 + 10 = 260	39 x 4 = (30 + 9) x 4 = 30 x 4 + 9 x 4 = 120 + 36 = 156
Halving and doubling		
65 x 8 = 65 x 2 x 4 = 130 x 4 = 520	24 x 6 = 48 x 3 = 144	32 x 8 = 64 x 4 = 128 x 2 = 256

Let us practice 12.2

1. 4 x 69 =
2. 85 x 9 =
3. 58 x 6 =
4. 51 x 8 =
5. 98 x 7 =

6. 79 x 8 =
7. 42 x 9 =
8. 75 x 5 =
9. 68 x 8 =
10. 96 x 8 =

12.3 Multiply a number by multiples of ten

Look at the trend in each column (top to bottom). Do you see how the number of zeros change has each number in the left column is multiplied by multiples of ten?

	X 10	x 100	X 1,000	X 10,000	X 100,000	X 1,000,000
1	10	100	1,000	10,000	100,000	1,000,000
2	20	200	2,000	20,000	200,000	2,000,000
3	30	300	3,000	30,000	300,000	3,000,000
4	40	400	4,000	40,000	400,000	4,000,000
5	50	500	5,000	50,000	500,000	5,000,000
6	60	600	6,000	60,000	600,000	6,000,000
7	70	700	7,000	70,000	700,000	7,000,000
8	80	800	8,000	80,000	800,000	8,000,000
9	90	900	9,000	90,000	900,000	9,000,000
10	100	1000	10,000	100,000	1,000,000	10,000,000

CHEETAH™
Connect to Higher Education, Electronic Tools, Aplication and Help

When we multiply a whole number by a multiply of 10, we add the zeros in the power of 10 to the right of the last digit in the whole number.

Other examples are:

- 31 x 100 becomes 3,100
- 654 x 10 becomes 6,540
- 204 x 1000 becomes 204,000
- 20 x 1,000,000 becomes 20,000,000

Let us practice 12.2

1. 1 x 10 =
2. 8 x 10 =
3. 14 x 100 =
4. 85 x 100 =
5. 39 x 1,000 =
6. 84 x 1,000 =
7. 48 x 100,000 =
8. 356 x 10 =
9. 234 x 1,000 =
10. 1,634 x 10 =

12.4 Identify and correct wrong answers in problems involving multiplication.

When we multiply, care must be taken to avoid these mistakes.

1. Times table mistakes e.g. 6 x 7 = 32

2. Doing the multiplication correctly but writing the wrong answer e.g. 3 x 2 = 5

3. Putting answers in the wrong position, particularly when multiplying in columns e.g. 42 x 12

$$\begin{array}{r} 42 \\ \times\ 12 \\ \hline 84 \\ \underline{42\ } \end{array}$$

The 2 in 42 is not in the correct place. It should be positioned under the 8, because you are multiplying by 10.

4. Not pulling the zero to hold the place. E.g. 16 x 32

$$\begin{array}{r} 16 \\ \times\ 32 \\ \hline 32 \\ \underline{48\ } \end{array}$$

5. Not carrying the tens e.g. 14 x 18

$$\begin{array}{r} 14 \\ \times\ 18 \\ \hline 82 \end{array}$$

The 3 from 32 is not carried.

6. Improper adding e.g. 89 x 7

$$\begin{array}{r} 89 \\ \times\ 79 \\ \hline 801 \\ \underline{623\ } \\ 6031 \end{array}$$

Let us practice 12.4

Find the multiplication errors in the problems.

1.	576 x 7 ————— 4052	2.	576 x 4 ————— 2084	3.	389 x 5 ————— 2055
4.	896 x 8 ————— 7078	5.	536 x 9 ————— 4844	6.	102 x 10 ————— 102
7.	4598 x 12 ————— 11086 5498 ————— 16574	8.	4598 x 43 ————— 14894 193720 ————— 207514	9.	3579 x 58 ————— 244632 178950 ————— 443582

12.5 Discover, memorize and recall all multiplication facts up to at least 12 x 12 = 144.

Let us practice 12.5

Complete the multiplication table by multiplying the column by the rows and putting the answers in the correct space.

Times	1	2	3	4	5	6	7	8	9	10	11	12
1												
2												
3												
4												
5												
6												
7												
8												
9												
10												
11												
12												

12.6 Differentiate between the use of addition and multiplication, subtraction and division in problem situations involving whole numbers.

The four operations in mathematics are addition (+), subtractions (-), multiplication (x) and division (÷). We use different keywords to tell which operation to use when solving problems.

Key words	
Addition	**Subtraction**
sum, add or altogether, total	difference, take away or subtract
Multiplication	**Division**
product, multiply or times	share, divide or go into.

Let us practice 12.6

Solve the problems below.

1. From the sum of 12 and 20, take their6differences.
2. To the sum of 20 and 4, add their differences.
3. To the product of 12 and 8, add their sum.
4. From the product of 5 and 4, take their sum.
5. To the difference of 12 and 8, add their product.
6. How many times can the difference of 20 and 30 go into their product?
7. Divide the product of 10 and 20 by their difference.
8. Divide the product of 20 and 5 by their sum.
9. Divide the sum of 20 and 40 by their difference.
10. Allana had 4 times as many pencils as John. If Allana had 48 pencils, how many did John have?
11. If Shawn has 6 times as many as Kimberly and Kimberly has 12 pens, how many pens does Shawn have?
12. Thadius as 8 marbles. Jevaughn has two times as many as Thadius and Ackeem has 3 times as many as Jevaughn. How many marbles do they have altogether?

12.7 Select data relevant to a problem when finding its solution.

Not all the data given or available in a mathematics problem is useful to solving the problem. Here, the useful information is said to be relevant to the problem. It is used to solve the problem. The information that is not useful is called irrelevant to the problem. Irrelevant information is not used in solving the problem.

Example: The road is 40 km long. Tom drove for 16 km and John drove for 18 km. How many km did they drive altogether?

Relevant information	Irrelevant Information
Tom drove for 16 km. John drove for 18 km. How many km did they drive altogether?	The road is 40 km long.
Solution: km driven altogether = 16 km + 18 km = 34 km	

Let us practice 12.7

For questions 1 to 3 insert the relevant and irrelevant information in the box, then solve the problem.

1. In 2021 a school had 736 students. In 2022 the school had 840 students. In 2023, the school had 785 students. How many students left the school between the years 2022 and 2023?

Relevant information	Irrelevant Information

2. A student needed to save $150 for the week. He saved $60 on Monday and $35 on Tuesday and Wednesday. How much did he save in all?

Relevant information	Irrelevant Information

3. Mikey has 6 puppies. His sister has 3 puppies and 10 kittens. His sister gave 6 kittens to the neighbours. How many puppies did they have in total?

Relevant information	Irrelevant Information

For questions 4 to 6, ignore the irrelevant information. Use the relevant information to solve the problem.

4. Mariana read 8 pages of a book. Janna read 6 pages of the book. Elaine read two times the amount that Mariana read. How many pages did Elaine read?

_____

5. Anya found 27 smooth stones at the river. Briana found 16 smooth stones. Janna lost 18 of the 20 smooth stones she found. How many stones does Janna have left?

_____

6. Look at the diagram of the playground. What information is relevant to find the perimeter of the playground? Calculate the perimeter.

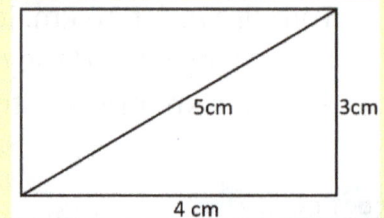

5cm 3cm

4 cm

_____

12.8 Identify and solve two-step problems. Write mathematical sentences for a two-step problem. Estimate and check answers to computations/problems.

When doing two step problems at least two mathematical problems are solved to reach the answer.

Example: Pam bought a dress for $5000 and sold it for $6000. She used $100 from what she earned to buy ice-cream. How much money does she have from what she earned?

Step 1.	Step 2.
How much did Pam earn = $6000 - $5000 = $1000	How much Pam has from what she earned = $1000 - $100 = $900

Let us practice 12.8

1. Mr. Scott bought two lunches: one for $850 and the other for $900 at a local restaurant. How much change would he receive if he paid $2000?

Step 1.	Step 2.

2. The wholesales downtown sell one notebook for $120. Thia buys 5 notebooks and sells each for $125 to the students at her school. How much money does she make?

Step 1.	Step 2.

3. Nathaniel's kite string is 45 m long. He lets go of 25 m more of string from the roll of string. If the roll of string is 100 m long, what is the length of string left on the roll?

Step 1.	Step 2.

4. Mrs. Dandy ordered five $500 lunches and fifteen $400 lunches for her class. How much money should she pay the canteen for the lunches?

Step 1.	Step 2.

5. Ms. Warren used her $400 gift voucher to purchase pens, pencils and notebooks at a bookstore. If each pen costs $50, each pencil costs $30 and each notebook costs $80, which combination of these items would allow Ms. Warren to use exactly all the money on her gift voucher?

   a. 3 pens, 2 pencils, 1 book   c. 4 pens, 3 books, 1 pencil

   b. 3 pencils, 3 pens, 2 books   d. 2 books, 3 pencils, 1 pen

6. Ms. Cummings asked her class to purchase 1 book for each of the 7 exam papers to be completed in grade 6. Each book costs $120 and $20 is taken off the cost of each book when 6 or more books are purchased. Find the cost of all 7 books.

Step 1.	Step 2.

Evaluation: Let us see how you did.

Learning outcome	No!	Working on it	Yes!
Did you get all the answers?	☹	😐	🙂
Did you get most questions right?	☹	😐	🙂
Did you retry the question(s) you got wrong?	☹	😐	🙂
Were you able to correct your wrong answers?	☹	😐	🙂
If not, did you seek help from others and/or review the chapter?	☹	😐	🙂

Colour the face that shows how you are doing.

How can I apply addition and subtraction to fractional numbers?

Before we begin, let's see what you know.

Prior learning:

✓ Distinguish between types of fractions.
✓ Know the place value of each digit in decimals.
✓ Add and subtract fractions with the same denominator up to 12ths.
✓ Model the addition and subtraction of fractions using fraction pieces or shading a grid.
✓ Subtract a proper fraction from whole numbers.

Key vocabulary

Check the words you understand:

☐ aquarium ☐ budget
☐ business ☐ cents
☐ common fractions
☐ cost
☐ decimal fractions
☐ decimal place(s)
☐ decimal point
☐ decimals ☐ fractions
☐ dollars
☐ entrepreneurship
☐ fractional numbers
☐ hundredths
☐ improper fractions
☐ loss ☐ mixed numbers
☐ money ☐ price
☐ profit ☐ taxes
☐ tenths

1. Tick the type(s) of fraction these fall under.

Numerals	0.25	$1\frac{1}{2}$	$\frac{8}{3}$	$\frac{3}{4}$
Proper fraction				
Improper fraction				
Mixed fraction				
Decimal fraction				

2. Which of these fractions are equivalent? $\frac{3}{4}$, $\frac{1}{2}$, $\frac{4}{8}$. _____

3. Shade the portions of the shape that show the fraction $\frac{5}{8}$.

4. Add or subtract the following fractions:

a. $\frac{3}{5} + \frac{4}{5} =$

b. $\frac{7}{12} + \frac{6}{12} =$

c. $\frac{3}{4} - \frac{1}{4} =$

d. $\frac{5}{8} - \frac{1}{8} =$

5. Write the fractions for the shaded grid problems.

a.

_____ + _____ = _____

b.

_____ - _____ = _____

6. Subtract these fractions:

a. $5 - \frac{3}{4} =$

b. $3 - \frac{1}{2} =$

c. $1 - \frac{5}{10} =$

d. $2 - \frac{3}{8} =$

If you know these, you should be able to learn what comes next.

13.1. Express fractional numbers with denominators 10 or 100 in decimal form or vice versa.
13.2 Write money in decimal form.
13.3 Complete sequence of fractional numbers in decimal form counting by tenths or hundredths.
13.4 Compute with decimals, including dollars and cents, using the four basic operations.
13.5 Investigate the base ten place value system when it is extended to show tenths and hundredths.
13.6 Add and subtract decimal fractions (including money).   13.7 Name whole numbers as fractions.
13.8 Solve real world problems involving the addition or subtraction of fractions with like denominators.

13.1 Express fractional numbers with denominators 10 or 100 in decimal form or vice versa.

A **fraction** that has a power of 10 as its denominator is called a **decimal fraction**. Examples: $\frac{7}{10}, \frac{21}{100}$

Remember that every whole number has a **decimal point** to the right, but we don't really write it. So, 7 is 7.0 and 21 is 21.0

For example: Let us convert $\frac{7}{10}$ to a **decimal**.

$$\frac{7}{10} = \frac{7.0}{10}$$

> When we divide by powers of 10, the decimal point in the answer is displaced to the left based on the number of zeros in the power of 10.

Divide denominator and numerator by 10. The denominator cancels to 1. The decimal point in the numerator moves 1 space because 10 has 1 zero.

$$\frac{7.0 \div 10}{10 \div 10} = \frac{0.7}{1} = 0.7$$

> Divide the numerator and the denominator by 100. The denominator cancels to 1. For the numerator, we move the decimal point 2 spaces to the left because 100 has 2 zeros.

Convert $\frac{21}{100}$ to a decimal.

$$\frac{21.0 \div 100}{100 \div 100} = \frac{0.21}{1} = 0.21$$

Now vice versa. Let us change a decimal to a fraction.

$$\frac{0.75 \times 100}{1 \times 100} = \frac{75}{100} = \frac{3}{4}$$

> Here we multiply the numerator and the denominator by 100, moving the decimal point 2 spaces to the right. To make $\frac{75}{100}$ a smaller fraction, we divide the numerator and denominator by 25 to get $\frac{3}{4}$.
>
> Or $\frac{0.25}{100}$ become $\frac{25}{100} = \frac{1}{4}$
>
> To bring the decimal to a fraction you can place a 1 under the decimal point and a zero for every digit following the decimal point.

Examples

$0.4 = \frac{4}{10}$	$0.04 = \frac{4}{100}$
$0.63 = \frac{63}{100}$	$0.801 = \frac{801}{1000}$
$7.5 = \frac{75}{10}$	$1.03 = \frac{103}{100}$

To convert a decimal to a proper fraction, we carry out the following steps.

Step 1: Look at the decimal and note the number of decimal places.

For example, 0.75 has 2 decimal places.

Step 2: Put the decimal over 2 powers of 10 (10 for each decimal place).

So, $0.75 = \frac{75}{100}$

Step 3: Cancel down. In this case $\frac{75}{100} = \frac{3}{4}$

Let us practice 13.1

1. Convert to decimal fraction.

a. $\frac{1}{10}$ =   b. $\frac{1}{100}$ =   c. $\frac{11}{100}$ =   d. $\frac{16}{100}$ =   e. $\frac{7}{10}$ =

f. $\frac{8}{10}$ =   g. $\frac{55}{100}$ =   h. $\frac{99}{1000}$ =   i. $\frac{9}{100}$ =   j. $\frac{3}{100}$ =

2. Convert decimals to proper fractions.

a. 0.2 =   b. 0.68 =   c. 0.4 =   d. 0.79 =   e. 0.025 =

f. 0.45 =   g. 0.8 =   h. 0.75 =   i. 0.601 =   j. 0.125 =

13.2 Write money in decimal form.

Note: ¢ is the **cents** sign and $ is the **dollar** sign.

100¢ = $1. When converting cents to dollars, we divide by 100.

> Note that the decimal point is moved 2 places to the right.

1. 40¢ = $\frac{40}{100}$ = $0.40   2. 250¢ = $\frac{250}{100}$ = $2.50 or $2 and 50¢

When converting dollars to cents, we multiply by 100.

1. $0.4 = 0.4 x 100 = 40¢   2. $1 and 20¢ = $1.20 = 1.20 x 100 = 120¢

Let us practice 13.2

1. Convert cents to dollars and cents.

a. 145¢ =   b. 340¢ =   c. 235¢ =   d. 175¢ =   e. 65¢ =

f. 845¢ =   g. 530¢ =   h. 95¢ =   i. 750¢ =   j. 1234¢ =

2. Convert dollars to cents.

a. $3.50 =   b. $0.50 =   c. $2.36 =   d. $5.50 =   e. $4.90 =

f. $12.30 =   g. $8.02 =   h. $6.95 =   i. $7.42 =   j. $25.31 =

13.3 Compute with decimals, including dollars and cents, using the four basic operations.

The 4 basic operations are addition, subtraction, multiplication and division.

Addition	Subtraction
Add the decimal numbers 2.46, 0.213, 0.064 and 21.23.	Take away the decimal numbers 17.89 from 23.52.

Addition

Add the decimal numbers 2.46, 0.213, 0.064 and 21.23.

```
  2.460
  0.213
  0.064
+21.230
-------
 23.967
```

Add the monies $65.45 + $24.35

```
  $64.45
 +$24.35
 -------
  $89.80
```

> Keep decimal points under one another. Add from right to left.

Subtraction

Take away the decimal numbers 17.89 from 23.52.

$$\begin{array}{r} {}^1\!2\;{}^{12}\;{}^{14} \\ 2\,3\,.\,5\,2 \\ -1\,7\,.\,8\,9 \\ \hline 5\,.\,6\,3 \end{array}$$

Adrian bought a strawberry sweetie from Thia for $6.23. He then sold the candy to his classmate for $18.35. What was the difference between his buying **price** and his selling price?

```
  $18.35
 - $6.23
 -------
  $12.12
```

> Keep decimal points under one another. Subtract from left to right.

Multiplication

Ignore the decimal. Multiply normally, then put in the decimal places starting on the right.

```
  16.43
 x  4.2
 ------
   3286
 +65720
 ------
 69.006
```

Count the number of decimal places in the numbers to be multiplied. Move one space for each decimal place in the two numbers you are multiplying.

Division

If 18.9 was divided by 1.4, how many times would 1.4 divide? $\frac{18.9}{1.4}$ = ?

First, convert the denominator to a whole number. To do so, we multiply by 10.

$$1.4 \times 10 = 14$$

Next, multiply the numerator by the same number you did to the denominator:

$$18.9 \times 10 = 189$$

Therefore, we have $\frac{189}{14}$.

```
        13.5
   14 )189
       -14↓
        49
       -42
         70
        -70
          0
```

So $\frac{18.9}{1.4}$ = 13.5 times

Let us practice 13.3

1. Add the following:

a. $26.85 + $0.93 + $3.92 + $0.06

b. 0.002 + 0.061 + 0.358 + 5.31

c. 21.68 g + 0.23 g + 9.07 g + 0.17 g

d. $12.47 + $0.03 + $6.92 + $19.57

e. 4.96 m + 2.04 m + 12.6 m + 0.08 m

f. 26.931 + 5.81 + 0.06 + 4.02

g. 42.6 °C + 0.3 °C + 5.2 °C + 16.9 °C

h. 75.964 + 0.023 + 5.861 + 0.007

i. 51.832 + 12.902 + 0.035 + 3.005

2. Subtract the following:

a.	b.	c.
$28.26 − $19.37	$40.37 − $38.59	648¢ − 596¢
Answer = $_____	Answer = _____¢	Answer = $_____, _____¢
d.	e.	f.
$94.06 − $85.79	$10.06 − $0.99	$2.67 − $1.98
Answer = $_____	Answer = $_____	Answer = $ _____, _____¢
g.	h.	i.
$46.78 − $37.89	$36.85 − $24.97	$88.15 − $79.68
Answer = $_____	Answer = $_____	Answer = $_____

CHEETAH™
Connect to Higher Education, Electronic Tools, Aplication and Help

j.	k.	l.
$13.886 - $12.897 Answer = $_____	$30.000 - $12.897 Answer = $_____	$1.99 - $0.99 Answer = _____ ¢

3. Multiply the following:

a. $4.90 x $3.40 = $4.9 x $3.4	b. 5.8 x 2.3	c. 1.23 x 3.2
d. 5.68 x 0.05	e. $12.35 x $4.5	f. 85.72 x 8.7
g. $54.65 x $39	h. 6.8 x 0.09	i. Find the product of 14.06 and 5.4.

4. Trevor bought a chocolate ball for $25 and 78¢. Later he bought 59 more chocolate balls for the same price. How much did Trevor pay for all the chocolate balls bought?
_____

5. Divide the following:

a. $19.20 ÷ $0.60	b. $15.35. ÷ $0.50	c. 36.6 ÷ 1.2

CHEETAH
Connect to Higher Education, Electronic Tools, Aplication and Help

d. 48.6 ÷ 0.6	e. $68.40 ÷ $1.20 =	f. 24.66 ÷ 0.9
g. 40.1 ÷ 1.3	h. 13.13 ÷ 1.3	i. 21.49 ÷ 0.7

6. Use addition and subtraction to solve these problems:

a. Add sixty-three dollars and ninety-five cents to thirty-eight dollars and sixty-eight cents.		b. Tom saved $138.67 and his dad gave him $149.36. How much does he have now?	
c. Fiona and her brother, Nicholas, bought ice-cream. She spent $154.67 and Nicholas spent $168.53. How much did they spend in all?		d. Five plums were bought at 76 cents each. How much was spent in dollars and cents?	
e. Three books were bought; one for $360.84 and two books for $140.75 each. How much was spent for all the books?		f. Mary had $294.32 and lent her friend $175.94. How much does she have left?	
g. Ann-Marie had $565.80 in her savings. Her mother borrowed $487.95. How much did she have left in her savings?		h. Timone had $916. He paid $568 for his lunch. How much did he have left?	
i. A vendor was owed $867.90 by a customer. The customer paid $678.50. How much money was left to be paid to the vendor?		j. Kristina has $865.86. Jada has $678.95. By how much is Kristina's money greater than Jada's?	

7. Use multiplication and division to solve these problems.

a. If one bag of sugar costs $455.16 with the tax included, what is the cost of 8 similar bags?		b. Eighteen students each gave five hundred and fifty-six dollars and fifty cents to a class party. How much money was collected?	
c. Find the cost of 8 apples at $248.13 each.		d. The taxi costs $235.50 per km. What is the fare for travelling 9 km?	
e. A typist typed 12 pages at $150.34 for each page. How much was she paid?		f. Divide $348.75 by 6.	
g. Share $985.05 equally among 9 children.		h. A bill for $8,650 was paid equally by 7 women. How much did each pay?	
i. If $ 29,780.64 was spent on 12 dictionaries, how much did each dictionary cost?		j. $1,240.00 was paid for one day at $155 per hour. How many hours were worked for the day? Write your answer in decimal form.	

13.4 Investigate the base ten place value system when it is extended to show tenths and hundredths.

Let's look at three hundred and ninety, point five-nine-six (390.596).

When expanded: $390.596 = 3 \times 100 + 9 \times 10 + 0 \times 1 + \frac{5}{10} + \frac{9}{100} + \frac{6}{1000}$

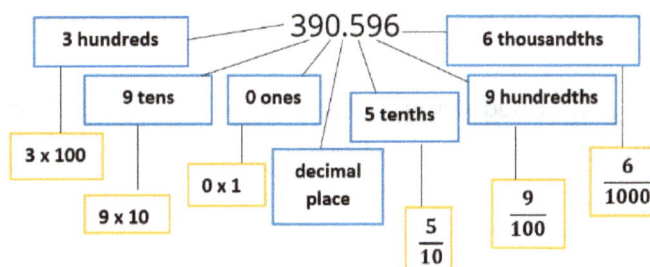

Let us practice 13.4

Expand the following decimal numbers.

1. 46.35 _____

2. 126.85 _____

3. 1.9 _____

4. 895.32 _____

5. 0.86 _____

13.5 Add and subtract decimal fractions (including money).

Example: Solve the following problem.

36.89 + 47.96 – 25.99

$$\begin{array}{r} \overset{1}{}\overset{1}{3}\,\overset{1}{6}.8\,9 \\ +\,4\,7.9\,6 \\ \hline 8\,4.8\,5 \\ -\,2\,5.9\,7 \\ \hline 5\,8.8\,8 \end{array}$$

Let us practice 13.5

1. Mrs. William's class needs $150 to purchase stickers. She collected $45.50, $16.84, $21.65 and $63.85. How much money is left to collected?

   _____

   _____

   _____

2. Anthony bought 2 pens at a cost of $124.45 each and 1 book costing $450.00. How much change should he receive out of $750.

   _____

   _____

   _____

3. If 1 kg of corn costs $85.5, find the cost of 3 kg of corn.

   _____

   _____

   _____

4. My mom had $787. She bought my sister a book for $135.67 and one for me for $451.84. How much does mom have left?

   _____

   _____

   _____

5. There was $967.85 in the class funds. $126.34 was taken out to buy pens for the class and $346.87 was taken out for photocopying material for the class. How much money is left?

_____

_____

_____

13.6 Name whole numbers as fractions.

Two is a whole number. As a fraction, 2 can also be written as $1 + \frac{4}{4}$ or $1 + \frac{5}{5}$.
Changing a whole number to a fraction is useful in problems involving mixed numbers. It is also useful in problems with whole numbers and proper fractions.

Example 1: $2 - \frac{3}{4} = 1 + \frac{4}{4} - \frac{3}{4}$

$= 1 + \frac{1}{4} = 1\frac{1}{4}$

So, $2 - \frac{3}{4} = 1\frac{1}{4}$

Example 2: $4\frac{3}{7} - 1\frac{5}{7} = 4 - 1 + \frac{3}{7} - \frac{5}{7}$

$3 - \frac{2}{7} = 3\frac{7}{7} - \frac{2}{7}$

So, $4\frac{3}{7} - \frac{5}{7} = 3\frac{5}{7}$

$= 3\frac{5}{7}$

Let us practice 13.6

Solve the following problems with proper fraction and whole numbers.

a. $3 - \frac{3}{4} =$	b. $3 + \frac{5}{6} =$	c. $3\frac{2}{8} - \frac{5}{8} =$	d. $5\frac{1}{3} - \frac{2}{3} =$

13.7 Solve real world problems involving the addition or subtraction of fractions with like denominators.

Let us practice 13.7

Real world problems.

1. Drake used the computer for a quarter of the day and Rihanna used it for a quarter of the day. What fraction of the day was the computer on?	
2. Justine ate $\frac{3}{8}$ of the cake and Britany ate $\frac{1}{8}$. What fraction of the cake was eaten?	

3. Neyo owed some money. If Neyo paid $\frac{3}{5}$ of the money, what fraction is not paid?

4. $\frac{5}{12}$ of the students in Class 402 are boys. What fraction are girls?

5. When the pizza was placed on the table to be shared. Ethan took $\frac{5}{8}$ and Jaden took $\frac{2}{8}$. Ethan was told to return $\frac{3}{8}$. What fraction of the pizza was now on the table.

6. Mama baked and sliced a pudding. Shantel took $\frac{3}{9}$ of the pudding slices and shanique took $\frac{5}{9}$. Shanique then gave $\frac{2}{9}$ to their little sister, Sasha. How much of the pudding slices were left for mama?

7. Trey, Ashanti, Keyshia and Mya decided to put money together to buy jerk chicken and festivals. What fraction can be used to share the food equally among them? Show how this fraction adds to make one whole.

13.8 Add or subtract mixed numbers, improper fractions and proper fractions with equal denominators.

$1\frac{3}{4}$ — mixed number — proper fraction — whole number	$1\frac{3}{4}$	$2\frac{1}{3}$ *can be converted to* $\frac{7}{3}$ Step 2: Add numerator to answer from step 1.
$1\frac{3}{4}$ = 1 + $\frac{3}{4}$ = $\frac{7}{4}$ mixed fraction / whole number / proper fraction / improper fraction		Step 1: Multiply whole number by denominator.

Example 1: $4\frac{3}{4} = \frac{(4 \times 4) + 3}{4} = \frac{16 + 3}{4} = \frac{19}{4}$

Example 2: $2\frac{3}{7} = \frac{(2 \times 7) + 3}{7} = \frac{14 + 3}{7} = \frac{17}{7}$

CHEETAH
Connect to Higher Education, Electronic Tools, Aplication and Help

13.9 Convert a mixed number to an improper fraction and vice versa.

Let us practice 13.8 & 13.9

Convert these mixed fractions to improper fractions.

1. $1\frac{2}{3}$	2. $2\frac{4}{5}$	3. $8\frac{1}{9}$
4. $5\frac{7}{8}$	5. $4\frac{4}{5}$	6. $6\frac{4}{10}$
7. $7\frac{5}{9}$	8. $9\frac{3}{5}$	9. $4\frac{7}{8}$
10. $6\frac{7}{12}$	11. $5\frac{2}{7}$	12. $1\frac{8}{11}$

13.10 Convert these improper fractions to mixed fractions

Example 1: Convert $\frac{12}{5}$ to a mixed fraction.	**Example 2:** Convert $\frac{17}{2}$ to a mixed fraction.	**Example 3:** Convert $\frac{43}{3}$ to a mixed fraction.
$\begin{array}{r} 2 \\ 5\overline{)12} \\ -10 \\ \hline 2 \end{array}$ ↑ remainder	$\begin{array}{r} 8 \\ 2\overline{)17} \\ -16 \\ \hline 1 \end{array}$	$\begin{array}{r} 14 \\ 3\overline{)43} \\ -3 \\ \hline 13 \\ -12 \\ \hline 1 \end{array}$
Answer = 2 Rem. 2 Or Ans $\frac{Rem}{Divisor}$ = $2\frac{2}{5}$	Answer = $8\frac{1}{2}$	Answer = $14\frac{1}{3}$

Let us practice 13.10

Convert these improper fractions to mixed fractions.

1. $\frac{3}{2}$	2. $\frac{7}{2}$	3. $\frac{9}{4}$

CHEETAH
Connect to Higher Education, Electronic Tools, Aplication and Help

4. $\frac{13}{5}$	5. $\frac{5}{4}$	6. $\frac{24}{7}$
7. $\frac{17}{9}$	8. $\frac{28}{5}$	9. $\frac{43}{3}$

13.11 Addition of mixed and improper fractions

Example 1: $\frac{1}{2} + 3\frac{1}{2}$

$= \frac{1}{2} + \frac{7}{2}$

$= \frac{8}{2}$ or 4

Example 2: $2\frac{3}{4} + 1\frac{3}{4}$

$= \frac{11}{4} + \frac{7}{4}$

$= \frac{18}{4}$ or $4\frac{2}{4}$ or $4\frac{1}{2}$

Let us practice 13.11

Solve the following.

1. $3\frac{2}{3} + \frac{2}{3}$	2. $1\frac{2}{7} + 4\frac{5}{7}$	3. $2\frac{1}{4} + 5\frac{1}{4}$
4. $7\frac{5}{6} + 4\frac{5}{6}$	5. $3\frac{4}{5} + 6\frac{1}{5}$	6. $3\frac{3}{9} + 4\frac{8}{9}$
7. $7\frac{3}{10} + 5\frac{9}{10}$	8. $1\frac{1}{2} + 3\frac{1}{2} + 7\frac{1}{2}$	9. $4\frac{1}{4} + 5\frac{3}{4} + 6\frac{2}{4}$

13.12 Subtraction mixed and improper fractions

Example 1: $3\frac{4}{5} - 1\frac{3}{5}$

$= \frac{19}{5} - \frac{8}{5}$

$= \frac{11}{5}$ or $2\frac{1}{5}$

Example 2: $4\frac{2}{5} - 1\frac{3}{5}$

$= \frac{22}{5} - \frac{8}{5}$

$= \frac{14}{5}$ or $2\frac{4}{5}$

CHEETAH
Connect to Higher Education, Electronic Tools, Aplication and Help

Let us practice 13.12

Solve the following.

1. $4\frac{3}{5} - 1\frac{2}{5}$	2. $6\frac{4}{7} - 2\frac{3}{7}$	3. $8\frac{5}{6} - 2\frac{1}{6}$
4. $3\frac{7}{9} - 1\frac{8}{9}$	5. $7\frac{5}{8} - 2\frac{7}{8}$	6. $4\frac{6}{11} - 3\frac{4}{11}$
7. $8\frac{3}{10} - 6\frac{7}{10}$	8. $5\frac{3}{5} - 2\frac{1}{5}$	9. $7\frac{1}{3} - 4\frac{2}{3}$
10. $2\frac{3}{7} - 1\frac{4}{7}$	11. $7\frac{3}{5} - 4\frac{4}{5}$	12. $6\frac{1}{3} - 5\frac{2}{3}$

Evaluation: Let us see how you did.

Learning outcome	No!	Working on it	Yes!
Did you get all the answers?	☹	😐	☺
Did you get most questions right?	☹	😐	☺
Did you retry the question(s) you got wrong?	☹	😐	☺
Were you able to correct your wrong answers?	☹	😐	☺
If not, did you seek help from others and/or review the chapter?	☹	😐	☺

Colour the face that shows how you are doing.

What is the difference between length and area and how are they measured?

Before we begin, let's see what you know.

Prior learning:
- ✓ Explain and use the term *perimeter*.
- ✓ Measure perimeter of polygons and various objects.

1. What is the perimeter of figure ABCDEF?

If you can answer this, you should be able to learn what comes next.

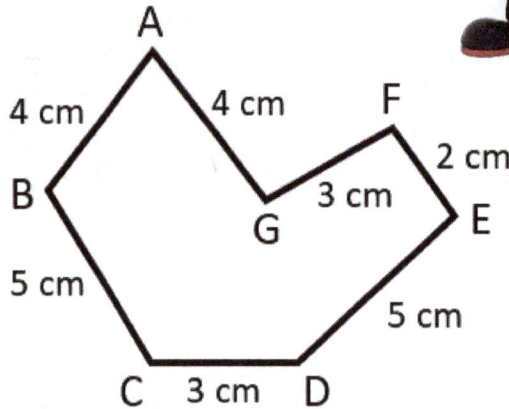

A

4 cm   4 cm   F

2 cm

B   3 cm

G   E

5 cm

5 cm

C   3 cm   D

14.1 Demonstrate an understanding of the difference between units of length and units of area.
14.2 Compare and contrast units of length and units of area.
14.3 Compute the perimeter of regular and irregular polygon, using units of measurement for length.
14.4 Find the area of various objects and figures.
14.5 Use unit squares or a centimetre grid to cover regions so as to determine their area.
14.6 Use a square grid (1 cm2) to find the area of any shape.

14.1 & 14.2 Demonstrate an understanding of the difference between units of length and units of area. Compare and contrast units of length and units of area.

Units of **length** can be measured using the metric units for metre: mm, cm, dm, m, dam, hm and km discussed in Chapter 5.

Area is measured from the multiplication (product) of two units of length (Example: cm x cm = **cm²**). This is why area is measured in units square (also called square units).

Length is the distance between any two points, but area is the space covered between lengths.	The area is the space covered by an object when it is placed on a flat surface. It is the space inside the lengths (bordered by the lengths).

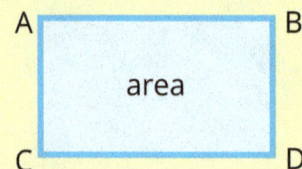

A ⊢————————————⊣ B   = length

A   B

area

C   D

Let us practice 14.1 & 14.2

1. From the units listed, insert the units used to measure lengths and those used to measure areas in the correct column.

List	Unit of length	Unit of area
a. mm, mm^2		
b. cm^2, km		
c. m, km^2		
d. m^2, cm		
e. mm^2, dm, m^2, hm, km^2		
f. cm^2, m, dm^2, m^2		
g. dam, hm^2, mm, cm^2, m^2, cm, mm^2, m		

2. Shashagaye inherited land from her grandmother. The land is flat, square and small enough for her to stand at one end and see the other ends. Which of the following would be most likely the unit used to measure the area of the land?

    a. km   b. m^2 c. km^2 d. m

14.3 Compute the perimeter of regular and irregular polygons using units of measurement for length.

A polygon is a two-dimensional closed-shaped figure that is flat and made with straight lines. It can be regular or irregular.

Examples of polygons include triangles, quadrilaterals, pentagons, hexagons, and many other shapes. See Chapter 9.

Perimeter is the distance around an object.

Example: Calculate the perimeter of Figure ABCDEFG.

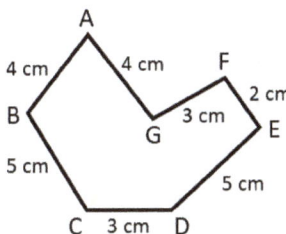

Perimeter = 2 cm + 5 cm + 7 cm + 6 cm + 4 cm

Perimeter = 24 cm

Let us practice 14.3

1. Find the perimeter of these regular polygons.

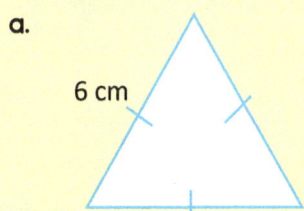

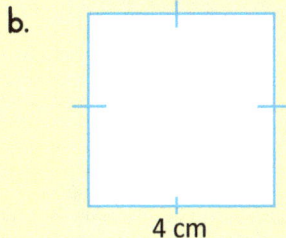

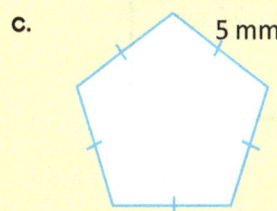

a.

6 cm

b.

4 cm

c.

5 mm

The markings on the sides mean the sides are equal in length.

d.

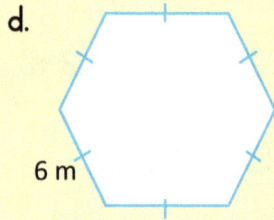

6 m

e.

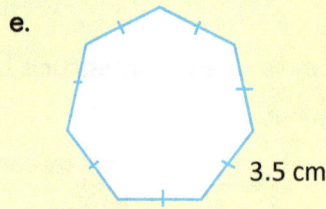

3.5 cm

2. Find the perimeter of these irregular polygons.

a.

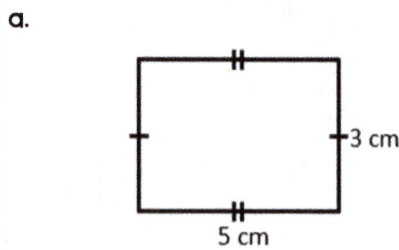

3 cm

5 cm

b.

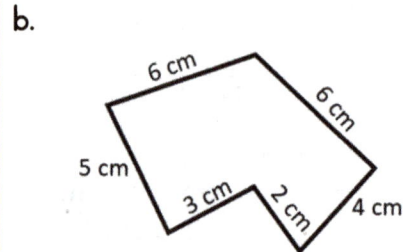

6 cm

6 cm

5 cm

3 cm

2 cm

4 cm

c.

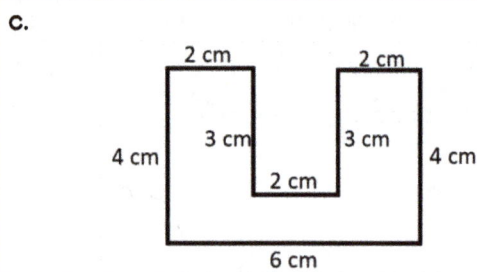

2 cm   2 cm

4 cm   3 cm   3 cm   4 cm

2 cm

6 cm

d.

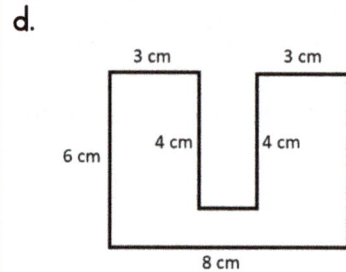

3 cm   3 cm

6 cm   4 cm   4 cm

8 cm

e.

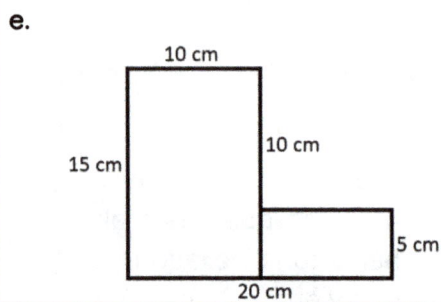

10 cm

10 cm

15 cm

5 cm

20 cm

f.

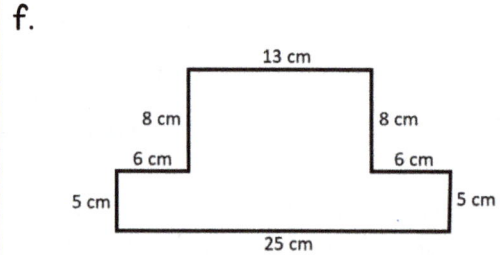

13 cm

8 cm   8 cm

6 cm   6 cm

5 cm   5 cm

25 cm

g. If the perimeter is 45 cm, find the value of d.

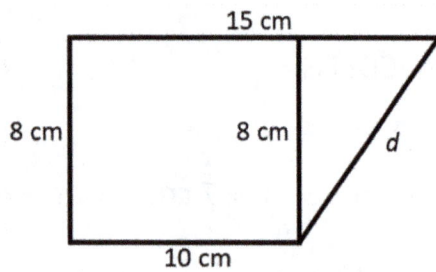

15 cm

8 cm   8 cm

d

10 cm

h.

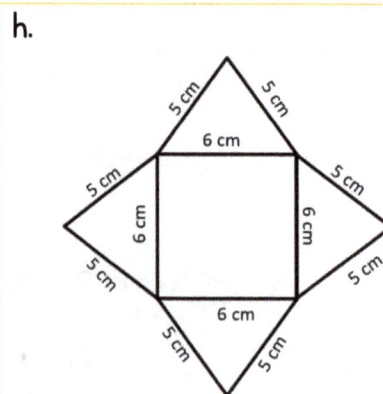

5 cm   5 cm

6 cm

5 cm   5 cm

6 cm   6 cm

6 cm

5 cm   5 cm

5 cm   5 cm

i.

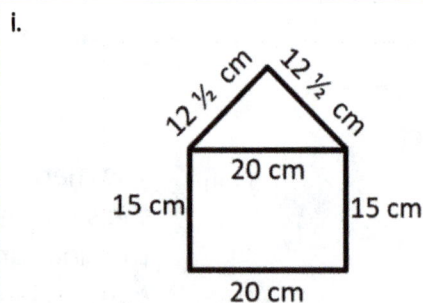

12 ½ cm   12 ½ cm

20 cm

15 cm   15 cm

20 cm

CHEETAH™
Connect to Higher Education, Electronic Tools, Aplication and Help

14.4 to 14.6 Find the area of various objects and shapes; Use unit squares or a centimetre grid to cover regions so as to determine their area; Use a square grid (1 cm squares) to find the area of any shape.

A rectangle is composed of two sides: length (L) and width (W). The length of a rectangle is the longest side, whereas the width is the shortest side. The width of a rectangle is sometimes referred to as the breadth (b).

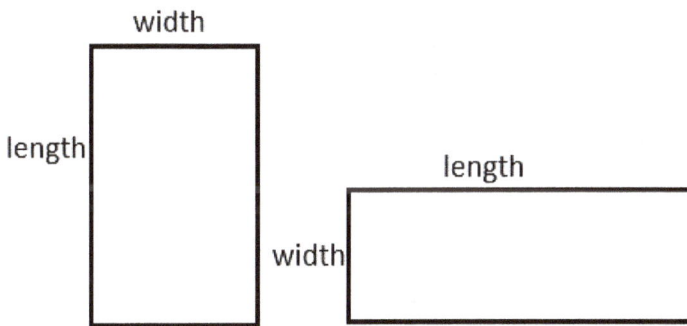

Two ways to find the area of a rectangular space

Area of a rectangle
= length x width
Or
Area of a rectangle = counted squares on a 1 cm x 1 cm grid.

The area of an object is the space covered by the object. This can be measured by drawing the object over a grid and counting the boxes in the grid. Each square is 1 cm long and 1 cm wide so the area of one small box is 1 cm². So, area is measured in unit square (Example: cm² or m²).

The area of a rectangular space may be calculated by (i) multiplying its length by its width or (ii) by placing the object or drawing on a **grid**, marking it out and counting the number of squares it covers.

Let us practice 14.4 to 14.6

1. Find the area covered by each grid by counting and multiplying rows by column. Look at the rectangles drawn on the grid and check to see that counting the grids covered is equal to multiplying the number of rows by the number of columns. The first grid is done for you.

Example:	a.	b.	c.
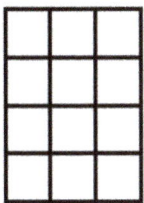		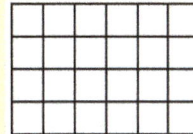	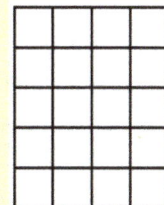

Count = 12 cm²
Area of each grid =
row x column
= 4 cm x 3 cm
= 12 cm²

Surface area

The surface area of a cuboid (box) is the area of all the faces (sides). Look at the cuboid below.

The cuboid has 6 faces: 1 in front, 1 back side, 1 on the left, 1 on the right, 1 at the top and 1 at the bottom. To find the surface area of this object, you must find the area of all 6 faces.

Sides of a cuboid
In a cuboid, the surface areas of the front and back sides are equal, the surface areas of the left and right sides are equal, while the surface areas of the top and bottom sides are equal.

Surface area of cuboid = area of front side + area of back side + area of left side + area of right side + area of top side + area of bottom side.

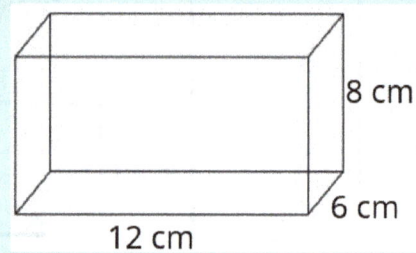

Area of front = 12 cm x 8 cm = 96 cm²
Area of back = 12 cm x 8 cm = 96 cm²
Area of left side = 8 cm x 6 cm = 48 cm²
Area of right side = 8 cm X 6 cm = 48 cm²
Area of top side = 12 cm x 6 cm = 72 cm²
Area of bottom side = 12 cm x 6 cm = 72 cm²

Surface area of cuboid = 96 cm² + 96 cm² + 48 cm² + 48 cm² + 72 cm² + 72 cm² = 432 cm². Answer is surface area of cuboid = 432 cm²

Using the grid, we count each small box as 1cm² and partial box as half or quarter box, as the case may be.

Example 1: Find the area of the figure if each square is 1 unit (length) x 1 unit (width).

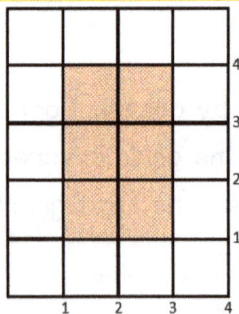

Calculating area
Area = row x column
   = 3 units x 2 units
   = (3 x 2) (unit x unit)
   = 6 units square
   = 6 units²

Finding area on a graph
To do this we count the number of squares and estimate spaces which are less than a whole space using fractions or decimals.

Example 2: Find the area of the octagon, if each square is 1 cm x 1 cm.

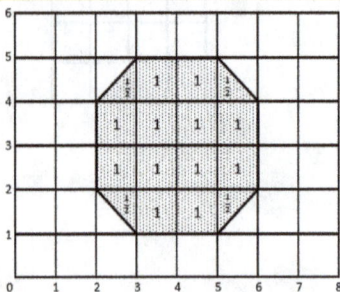

Finding area on a graph by counting the squares

Area = ½ + 1 +1 + ½ + 1 + 1 + 1 + 1 + 1 + 1+ 1 + 1 + ½ + 1 + 1 + ½

Area = ½ +2 + ½ + 8 + ½ + 2 + ½
   = 14 cm²

CHEETAH
Connect to Higher Education, Electronic Tools, Aplication and Help

Example 3: Draw the outline of a leaf on a grid to find its area by counting the squares

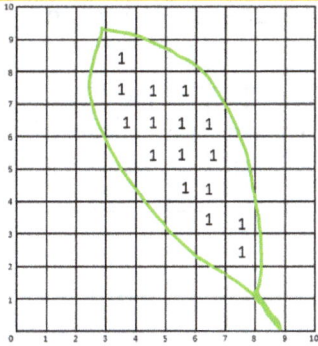

Find the area of a leaf using a graph by counting the squares, where each square is 1 cm x 1 cm.

Area = 16 full squares + 23 partially covered squares = 16 + 11 ½ = 27 ½

Area is approximately 27 ½ cm^2

Let us practice 14.4 to 14.6 continued

2. A garden was planted with four sets of crops.

a. What is the perimeter of the garden? _____

b. What units square would be used to measure the garden's area? _____

c. What is the area of the pepper plot?_____

d. Calculate the area of the corn section of the garden. _____

e. Which area is larger, the plot with the onions or tomatoes and by how much? _____

3. Find the area of the hexagon if each square is 1 cm by 1 cm.

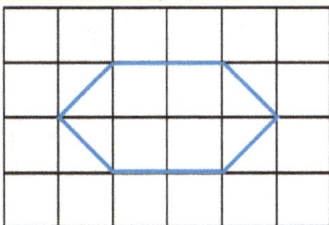

4. Find the area of this kite if each square is 1 cm by 1 cm.

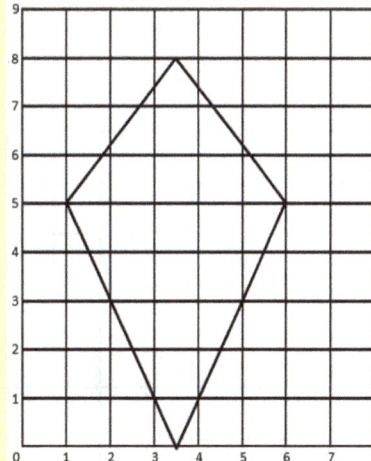

5. Find the area of this figure given that each square is 1 cm by 1 cm.

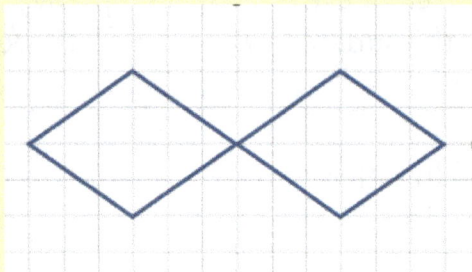

6. Find the area of the outline of one side of a building on the grid, where each square is 1 m by 1 m.

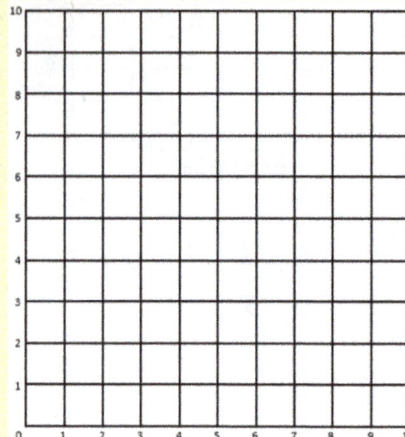

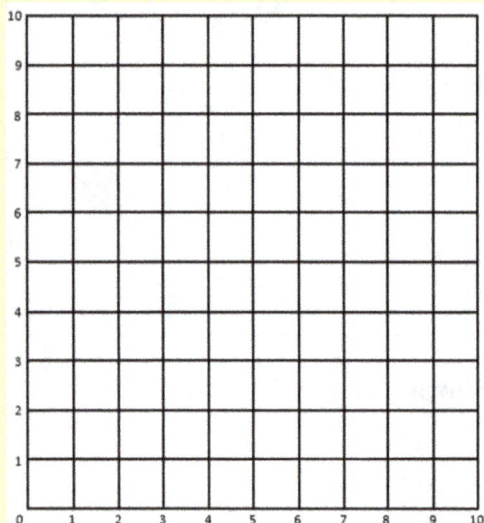

7. Draw the outline of one side of a building on the grid, where each square is 1 m by 1 m.

8. Find the perimeter and area of this shape, if each square is 1 cm x 1 cm.

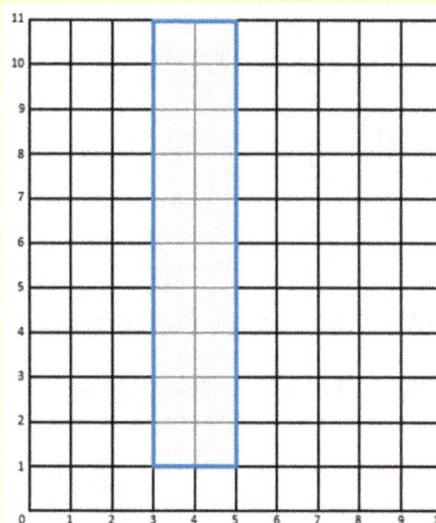

9. Find the area of a circle on the grid, if each square is 1 cm x 1 cm.

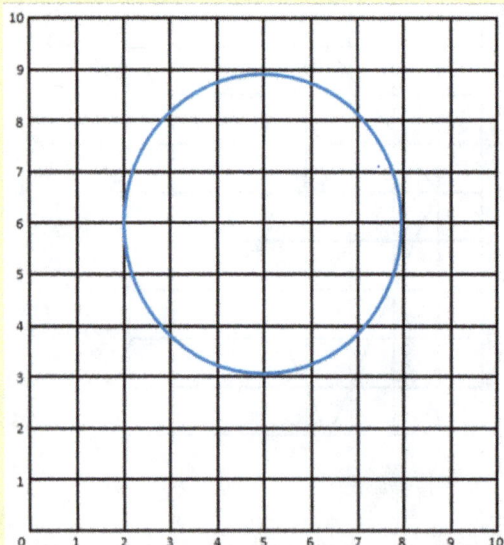

10. Trace the outline of a small leaf on the grid. Find the area of the outline, if each square is 1 cm x 1 cm.

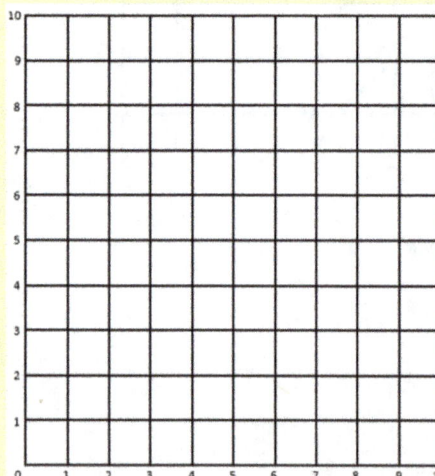

11. On graph paper, trace the outline of a leaf and find its area if each square is 1 cm x 1 cm.

12. On graph paper, trace the outline of the bottom of a bottle and find its area if each square is 1 cm x 1 cm.

13. On graph paper, trace the outline of a small ruler and find its area if each square is 1 cm x 1 cm.

14. Find the surface area of the cuboid without a cover.

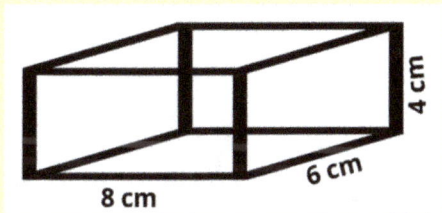

8 cm   6 cm   4 cm

15. Find the surface area of the cuboid with a cover.

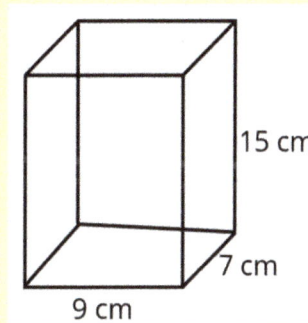

15 cm   7 cm   9 cm

Evaluation: Let us see how you did.

Learning outcome	No!	Working on it	Yes!
Did you get all the answers?	☹	😐	☺
Did you get most questions right?	☹	😐	☺
Did you retry the question(s) you got wrong?	☹	😐	☺
Were you able to correct your wrong answers?	☹	😐	☺
If not, did you seek help from others and/or review the chapter?	☹	😐	☺

Colour the face that shows how you are doing.

What are the things around us that have lines of symmetry?

Before we begin, let's see what you know.

Prior learning:
- ✓ Divide an object in two equal parts.
- ✓ Identify the circle, polygons and 3-D shapes.
- ✓ Identify the diameter of a circle.

Key vocabulary

Check the words you understand:

- ☐ angle
- ☐ diameter
- ☐ radius
- ☐ polygon

1. Draw an image of the orange to show it divided into 2 parts.

2. Name these shapes.

a.	b.	c.	d.
_____	_____	_____	_____
e.	f.	g.	h.
_____	_____	_____	_____

3. Select from the list of names and insert the correct name of these 3-dimensional (3D) shapes.

Image of shape	Name of shape	Description of shape
a.	_____	2 triangular bases and 3 rectangular sides
b.	_____	2 square bases and 4 rectangular sides
c.	_____	3-D box image with 6 square sides

List of names: square-based prism, cube and triangular-based prism.

4. State what you know about a circle.

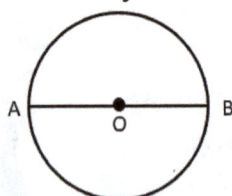

a. Name line AB: _____

b. Name line OB: _____

c. Is line OA = OB? _____

If you know these, you should be able to learn what comes next.

15.1   Associate symmetry with reflection.
15.2   Identify the mirror line of a reflection.
15.3   Identify the mirror line as being a line of symmetry.
15.4   Show the diameter of a circle as a line of symmetry.
15.5   Identify the possible lines of symmetry in geometric shapes and objects.

15.1 Associate symmetry with reflection

Look at an image of yourself in the mirror. What you see is your reflection. Do you notice that you (the object) and the image in the mirror are exact. If you move away from the mirror, the image moves away by the same distance. The image and its reflection are the same size and shape. The image is the same distance behind the mirror as the object is in front of the mirror.

The mirror is said to be a line of symmetry between the object and the image. The image is a reflection of the object in the mirror. An object has symmetry, if the object can be split into two mirror images. An example is a butterfly.

line of symmetry

Symmetric

line of symmetry

line of symmetry

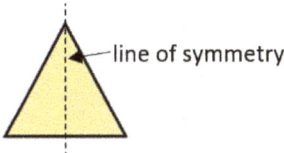

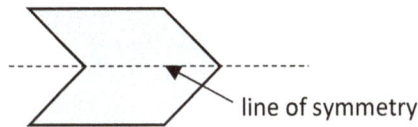

Not symmetric

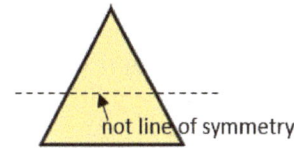

not line of symmetry

not line of symmetry

Let us practice 15.1

Colour the shapes that have symmetry along the line drawn.

1.	2.	3.	4.	5.
6.	7.	8.	9.	10.

CHEETAH™
Connect to Higher Education, Electronic Tools, Aplication and Help

11.	12.	13.	14.	15.

15.2 Identify the mirror line of a reflection.

Remember that the mirror line divides the figure exactly in two equal parts or shapes. The size and appearance of each part is identical.

Let us practice 15.2

If the mirror line is a line of symmetry, complete the image of the figure on the other side of the line of symmetry.

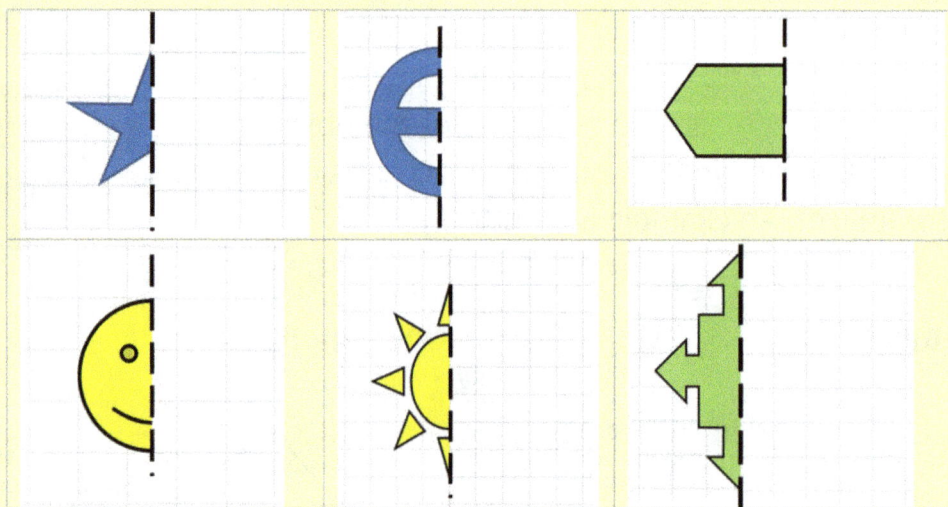

15.3 Identify the mirror line as being a line of symmetry

The mirror line is a line of symmetry.

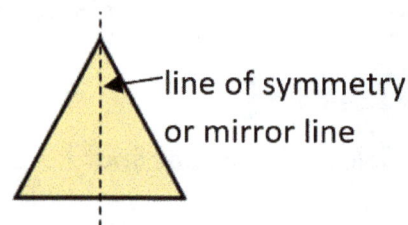

line of symmetry or mirror line

15.4 Show the diameter of a circle as a line of symmetry

A diameter is a line that stretches from one end of a circle to another through the centre (O). The diameter of a circle divides the circle into two equal halves. This means that the diameters of the circle are all lines of symmetry.

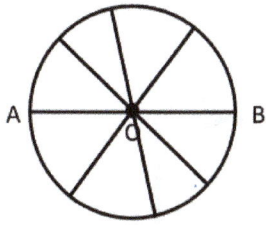

A circle has an infinite number of lines of symmetry. Each line cuts the circle in two halves.

Let us practice 15.4

Mark the centre of the circle. Draw 5 lines of symmetry through the circle to run from one side of the circle through the centre to end as it touches the circle on the other side. Measure each line and write one sentence about your measurement results.

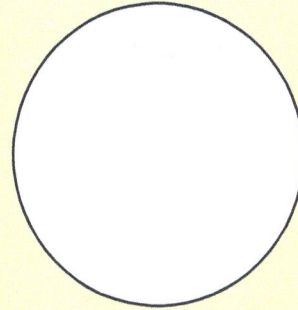

Sentence: _____

15.5 Identify the possible lines of symmetry in geometric shapes and objects.

	Examples
Some geometric shapes have no line of symmetry. There is no line you can draw through the shape to get two identical halves.	
Some geometric shapes have one line of symmetry.	
Others geometric shapes have more than one line of symmetry.	

Let us practice 15.5

1. Draw all the lines of symmetry seen in the objects. State if they have 0, 1 or 2 lines of symmetry.

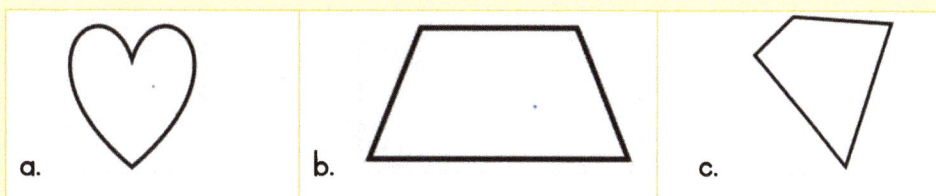

   a.   b.   c.

Connect to Higher Education, Electronic Tools, Aplication and Help

d. ⬭	e. △	f. 🏴
g.	h. ⟷	i. ⇒

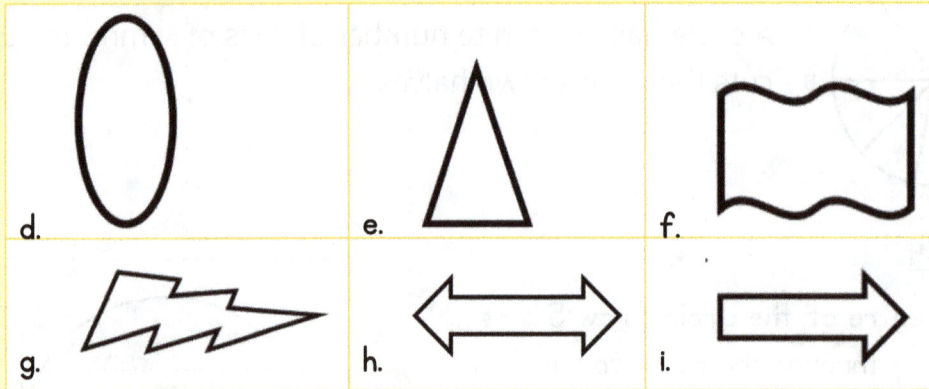

2. Look at the letters of the alphabet. Write whether each has 0, 1 or 2 lines of symmetry.

A	B	C	D
E	F	G	H
I	J	K	L
M	N	O	P
Q	R	S	T
U	V	W	X
Y	Z		

Evaluation: Let us see how you did.

Learning outcome	No!	Working on it	Yes!
Did you get all the answers?	☹	😐	🙂
Did you get most questions right?	☹	😐	🙂
Did you retry the question(s) you got wrong?	☹	😐	🙂
Were you able to correct your wrong answers?	☹	😐	🙂
If not, did you seek help from others and/or review the chapter?	☹	😐	🙂

Colour the face that shows how you are doing.

Focus question

TERM 2, UNIT 3

Chapter 16

What are the characteristics of geometric shapes in different orientations?

Before we begin, let's see what you know.

Prior learning:
- ✓ Identify rows and columns.
- ✓ Trace the path of an object.
- ✓ manipulate concrete objects, flip or slide.

1. In the soldier layout shown count and record the number of columns and the number of rows seen?

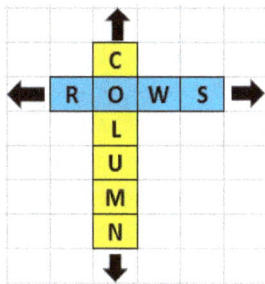

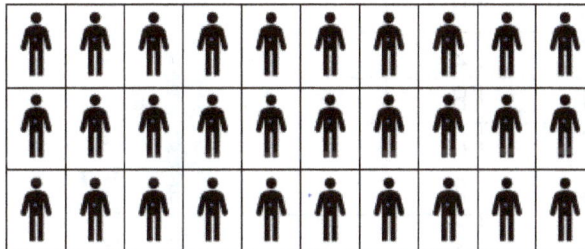

2.

rows:_____   columns: _____

3. Use a pencil to trace 5 pathways from start (S) to finish (F), without using oblique lines. Moving along horizontal (row) and vertical (column) lines only, colour and name the shortest path blue and the longest path red to travel from S to F.

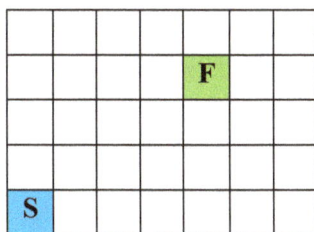

Count and state the number of boxes moved for the shortest and longest paths.

Shortest path: _____

Longest path: _____

If you know these, you should be able to learn what comes next.

Key vocabulary

Check the words you understand:

- ☐ columns
- ☐ congruent
- ☐ flip
- ☐ horizontal
- ☐ image
- ☐ location
- ☐ object
- ☐ orientation
- ☐ rows
- ☐ shape
- ☐ size
- ☐ slide
- ☐ vertical

16.1 Describe locations on a grid using columns and rows.
16.2 Make inferences about congruency when a shape or design is flipped, turned or slid.
16.3 Identify details in shapes and designs from different orientations and perspectives.

16.1 Describe locations on a grid using columns and rows.

Locations on a grid are described in **columns** and **rows**. For the grid on the next page, the rows are numbered and the columns are identified using letters.

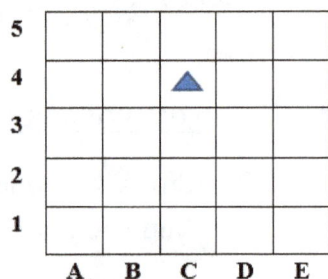

To state the location of the triangle on the grid, we identify which column and which row the triangle is in.

Example: The triangle is in column C and row 4.
The location of the triangle on the grid is C4

Let us practice 16.1

1. Identify the numbers in the following locations on the number grid given.

Location on number grid

Columns	Rows	Answer	Columns	Rows	Answer
A	1		C	9	
D	6		F	2	
C	4		J	8	
H	5		E	1	
J	7		B	5	
F	9		E	6	
G	2		A	3	
H	10		G	7	
I	3		D	10	
J	8		B	4	

Number grid

10	1	2	3	4	5	6	7	8	9	10
9	11	12	13	14	15	16	17	18	19	20
8	21	22	23	24	25	26	27	28	29	30
7	31	32	33	34	35	36	37	38	39	40
6	41	42	43	44	45	46	47	48	49	50
5	51	52	53	54	55	56	57	58	59	60
4	61	62	63	64	65	66	67	68	69	70
3	71	72	73	74	75	76	77	78	79	80
2	81	82	83	84	85	86	87	88	89	90
1	91	92	93	94	95	96	97	98	99	100
	A	B	C	D	E	F	G	H	I	J

2. State the positions in columns and rows for the locations of these numbers on the grid.

Locations of numbers on grid

Number	Location	Number	Location
3		22	
84		16	
53		31	
79		100	
46		36	
58		17	
12		28	
33		77	
94		82	
1		50	

Number grid

10	1	2	3	4	5	6	7	8	9	10
9	11	12	13	14	15	16	17	18	19	20
8	21	22	23	24	25	26	27	28	29	30
7	31	32	33	34	35	36	37	38	39	40
6	41	42	43	44	45	46	47	48	49	50
5	51	52	53	54	55	56	57	58	59	60
4	61	62	63	64	65	66	67	68	69	70
3	71	72	73	74	75	76	77	78	79	80
2	81	82	83	84	85	86	87	88	89	90
1	91	92	93	94	95	96	97	98	99	100
	A	B	C	D	E	F	G	H	I	J

16.2 Make inferences about congruency when a shape or design is flipped, turned or slid.

Congruency means exactly equal in shape and size. The **shape** and **size** should remain even when the **shape** is flipped, turned or slid. If you cut out shapes after they are flipped or turned and put them on top of each other and they fit exactly, then they are **congruent**.

Movement of congruent shapes	Examples
A **flip** is a reflection. This is done across a mirror line.	
A **turn** is a rotation. This is a spin around a point. Clockwise follows the same direction as a clock. Counter-clockwise moves in the opposite direction to a clock.	
A **slide** is a left to right or up to down shift on a flat surface. A slide is done in a straight line.	

Let us practice 16.2

1. Identify the type of movement shown for each congruent shape.

Movement of congruent shape on a grid		
 a. Movement: _____	 b. Movement: _____	 c. Movement: _____
 d. Movement: _____	 e. Movement: _____	 f. Movement: _____
 g. Movement: _____	 h. Movement: _____	 i. Movement: _____

2. Trace the path of an object to its image by identifying the number of units traveled horizontally, then the number of units traveled vertically. Move the image on the right to the one on the left.

Movement of a congruent shape on a grid

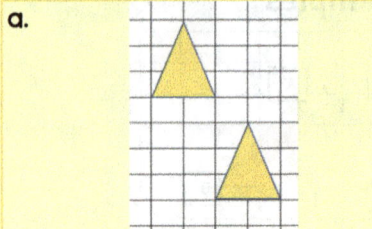

a.

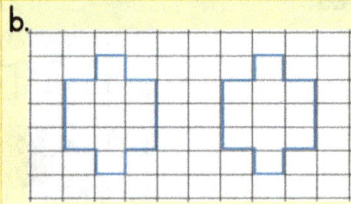

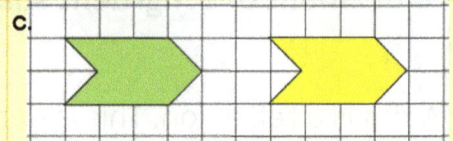

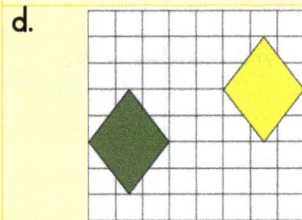

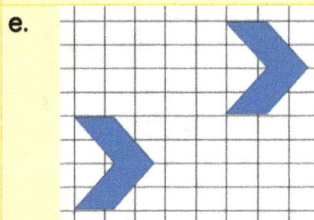

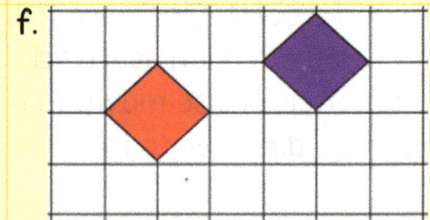

Horizontally: _____

Vertically: _____

b.

Horizontally: _____

Vertically: _____

c.

Horizontally: _____

Vertically: _____

d.

Horizontally: _____

Vertically: _____

e.

Horizontally: _____

Vertically: _____

f.

Horizontally: _____

Vertically: _____

3. Use the clues given to play the game below. Locate the object by stating the column followed by the row. Use each face only once.

Clues:

a. I have a triangular shaped nose. I am wearing a frown. I wear a hat, but no tie. Where am I located?

b. I am wearing a bow tie. I have a triangular-shaped nose. I have a crooked smile. Where am I located?

c. My eyes are triangular-shaped. I have a circular nose. I am wearing a smile. Where am I located?

d. I have 2 circular eyes. My nose is triangular shaped. I have a leaf. I am wearing a frown. Where am I located?

e. I am wearing a frown. My nose is heart shaped. I have a leaf hat. My eyes are stars. Where am I located?

Source: Extract from the National Standards Curriculum (NSC) Exploratory Core for Grade 4, 2018 published by the MOEYI Jamaica.

16.3 Identify details in shapes and designs from different orientations and perspectives.

Details in shapes and designs can be noted from different **orientations** and perspectives when shapes or designs are flipped or turned across a grid. Both the flip and turn can change the orientation. Given the original **image**, and the units travelled **horizontally** and **vertically**, we can draw the new location of the image.

Let us practice 16.3

1. Examine the different orientations of the shapes shown. In the boxes, insert the missing letter names for the vertices of each shape after the change.

Flip the rectangle ABCD across the dotted line shown.

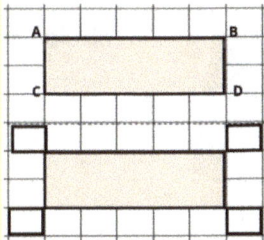

Flip the triangle XYZ across the line shown.

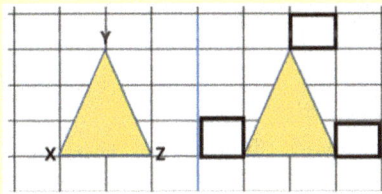

Turn around point Y of triangle XYZ.

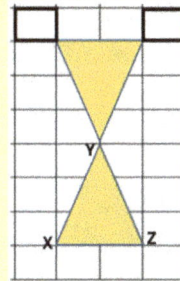

Rotate anti-clockwise about the blue dot.

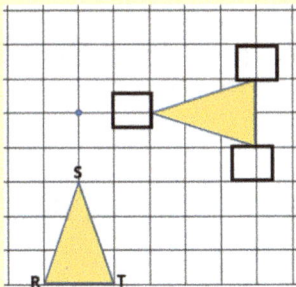

Turn around point P of triangle PQR.

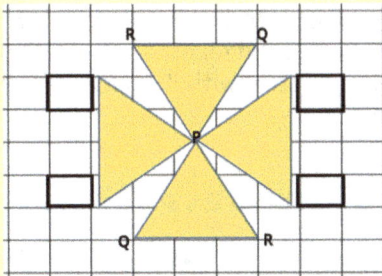

Slide 4 over to the right and then 2 up.

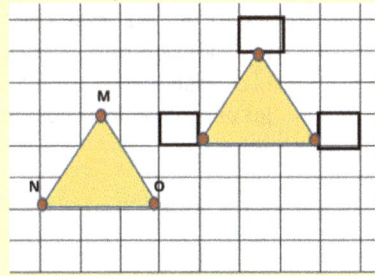

Evaluation: Let us see how you did.

Learning outcome	No!	Working on it	Yes!
Did you get all the answers?	☹	😐	🙂
Did you get most questions right?	☹	😐	🙂
Did you retry the question(s) you got wrong?	☹	😐	🙂
Were you able to correct your wrong answers?	☹	😐	🙂
If not, did you seek help from others and/or review the chapter?	☹	😐	🙂

Colour the face that shows how you are doing.

How do I make sense of different patterns?

Before we begin, let's see what you know.

Prior learning:
✔ Describe number patterns using terms such as, "one less" and "one more."

Key vocabulary

Check the words you understand:

☐ design
☐ expressions
☐ operations
☐ patterns
☐ positions
☐ prediction
☐ sequence
☐ tabular form

1. Look at these number patterns. What do you see in the number patterns?

Number patterns	What can you say about the pattern of each set of numbers?
(a) 1, 2, 3, 4, 5, 6, 7	
(b) 9, 8, 7, 6, 5, 4, 3	

If you know these, you should be able to learn what comes next.

🎯 17.1 From a patterning rule expressed in words, develop number patterns using any of the four arithmetic operations; Make predictions for numerical patterns.
17.2 Design and describe geometric patterns; Make predictions for geometric patterns.
17.3 Associate each term in a pattern with its position in the sequence and express this information in a tabular form.

17.1 From a patterning rule expressed in words, develop number patterns using any of the four arithmetic operations; Make predictions for numerical patterns.

Patterns can be seen everywhere around us. They are used to create **designs**. The elements in a pattern are mathematically linked. These elements can be a series of repeated numbers, shapes or even colours. Every pattern follows a rule. For each pattern, the number, shape or colour would repeat to follow the rule using one or more arithmetic **operations** (+, -, x or ÷). This repetition of the elements in a pattern allows us to see the rule.

Here is an example of different number patterns and the rules they follow. Look at the descriptions and patterns in the table and see if you come up with the same rules.

Description of the pattern	Pattern	Rule
Pattern of even numbers	2, 4, 6, 8, 10, 12...	Add 2 to each.
Pattern of odd numbers	1, 3, 5, 7, 9, 11 ...	List of odd numbers starting at 1.
Pattern of prime number	1, 2, 3, 5, 7, 11...	List of prime numbers starting at 1.

Description of the pattern	Pattern	Rule
Pattern of even numbers	14, 12, 10, 8, 6, ...	Subtract 2 from each number starting at 14.
Pattern multiples of 5	5, 10, 15, 20, 25,	Add 5 every time.
Pattern of factors	64, 32, 16, 8, 4 ...	Divided by 2 starting at 64.
Pattern of add then subtract	1, 4, 3, 6, 5, 8, ...	Add 3 to 1st term, then subtract one to next term, then add 3 and continue.

Let us practice 17.1

1. Write the rules and the next term for the patterns given.

	Pattern	Rule	Next term
a.	4, 8, 12, 16, 20, …		
b.	10, 20, 30, 40, 50, …		
c.	23, 21, 19, 17, 15, …		
d.	7, 14, 21, 28, …		
e.	18, 27, 36, 45, …		
f.	55555, 4444, 333, 22, …		
g.	32, 27, 22, 17, …		
h.	7, 11, 13, 17, 19, ….		
i.	97, 95, 93, 91, …		
j.	41, 45, 44, 48, 47….		

2. Write the first four members of the patterns from the rules given.

	Pattern		Rule
a.			List even numbers starting at 6.
b.			List multiples of 3 starting from 9.
c.			List factor of 12 in ascending order.
d.			List alternate odd numbers.
e.			List whole numbers greater than 4.
f.			List repeated division of 100 by 2.
g.			List squares of numbers 1 to 4.
h.			List whole numbers multiplied by 3 starting at 1.
i.			List of whole numbers decreasing by 3 from 15.
j.			List of whole numbers increasing by 7 from 18.

CHEETAH
Connect to **H**igher **E**ducation, **E**lectronic **T**ools, **A**plication and **H**elp

3. A list of numbers that follows a certain sequence is a pattern or number pattern. Which numbers come next in these number patterns?

A. Add the next number in the series	B. Add the next two numbers in the series
i. 2, 4, 6, ____.	i. 5. 10, 15, 20, ____,____.
ii. 1, 3, 5, 7, ____.	ii. 20, 16, 12, 8, ____,____.
iii. 1, 2, 4, 8, ____.	iii. 12, 18, 24, 30, ____,____.
iv. 3. 6, 9, 12, ____.	iv. 40, 35, 31, 28, ____,____.
v. 10, 15, 20, ____.	v. 32, 38, 43, 47, ____,____.
vi. 1, 5, 9, 13, ____.	vi. 1, 4, 9, 25, ____,____.
vii. 2, 5, 9, 14, ____.	vii. 1, 2, 5, 10, ____,____.
viii. 20, 16, 13, 11, ____.	viii. 19, 14, 10, 7, ____,____.
ix. 24, 19, 14, 9, ____.	ix. 30, 35, 29, 34, ____,____.
x. 20, 24, 23, 27, ____.	x. 32, 26, 21, 17, ____,____.

4. Which letter or group of letters is next in this pattern of letters of the alphabet?

i. A, C, E, G, ____.	vi. BA, CD, FE, HG, ____.
ii. Z, X, V, T, ____.	vii. G, H, L, P, ____.
iii. ZA, YB, ZC, WD, ____.	viii. W, S, O, K, ____.
iv. CD, GH, KL, NO, ____.	ix. BA, ED, HG, KJ, ____.
v. XW, TS, PO, LK, ____.	x. WA, WD, WG, WJ ____.

17.2 Design and describe geometric patterns; Make predictions for geometric patterns.

A geometric pattern is made by increasing/decreasing a number of smaller shapes kept together. In the geometric pattern, a series of shapes is drawn to follow a design. The number of smaller shapes usually follows a number pattern.

A number pattern can be seen in each geometric pattern.

Example: Geometric pattern of triangles

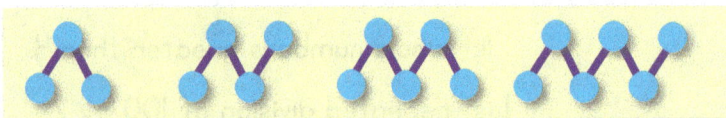

Image number: 1 2 3 4 5 6 7 8

Number of circles: 3   4   5   6   7   8   9   10

The circle pattern is described as increasing by 1.

What is the rule that tells you how many circles are in the next image?

Answer: Each image increases by a circle.

How many circles are in figure 8?

Answer: If the number of circles increases by 1 for each image and image 4 has 6 circles, then image 8 would have 4 more circles. Hence, image 8 has 10 circles.

CHEETAH
Connect to Higher Education, Electronic Tools, Aplication and Help

Within the geometric pattern, the arithmetic operations of multiplication and division can tell the next character or element in the series. If given three or more numbers in the **sequence**, you can find the unknown numbers in the pattern using multiplication and division operations. Geometric patterns are more complex sequences and have a growing pattern.

1. Complete the geometric pattern in the table.

Number of triangles	1	2	3	4	5	6	7	8	9	10
Number of edges	3	6								

Describe the pattern seen in words:

_____

2. Draw the shape that comes next in the geometric pattern. Describe the geometric pattern by writing the number pattern below each.

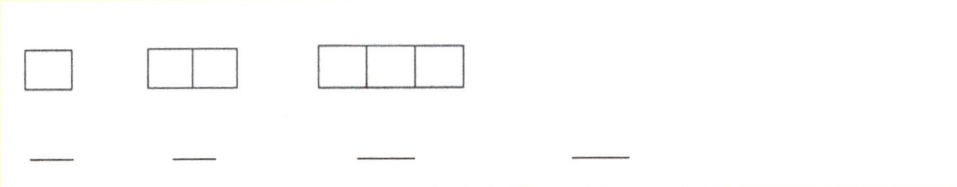

3. Complete by drawing the next pattern in the set:

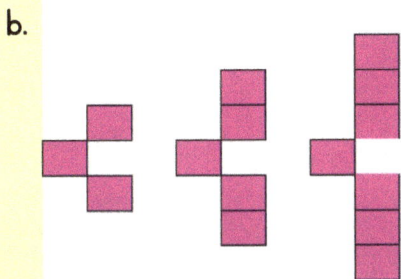

   a.

   b.

   c.

d.

e.

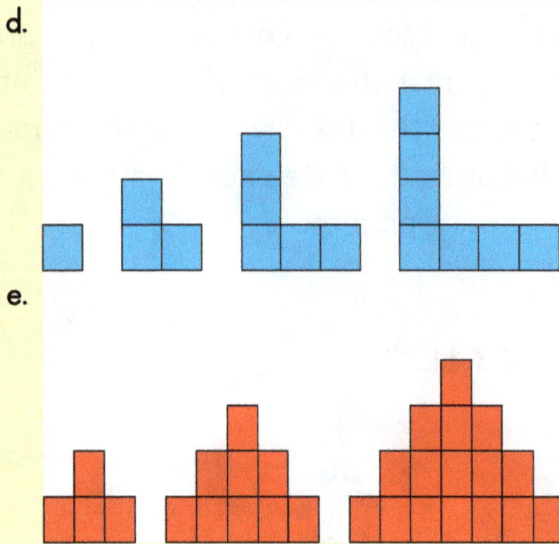

4. Draw and colour the next shape in this pattern of shapes.

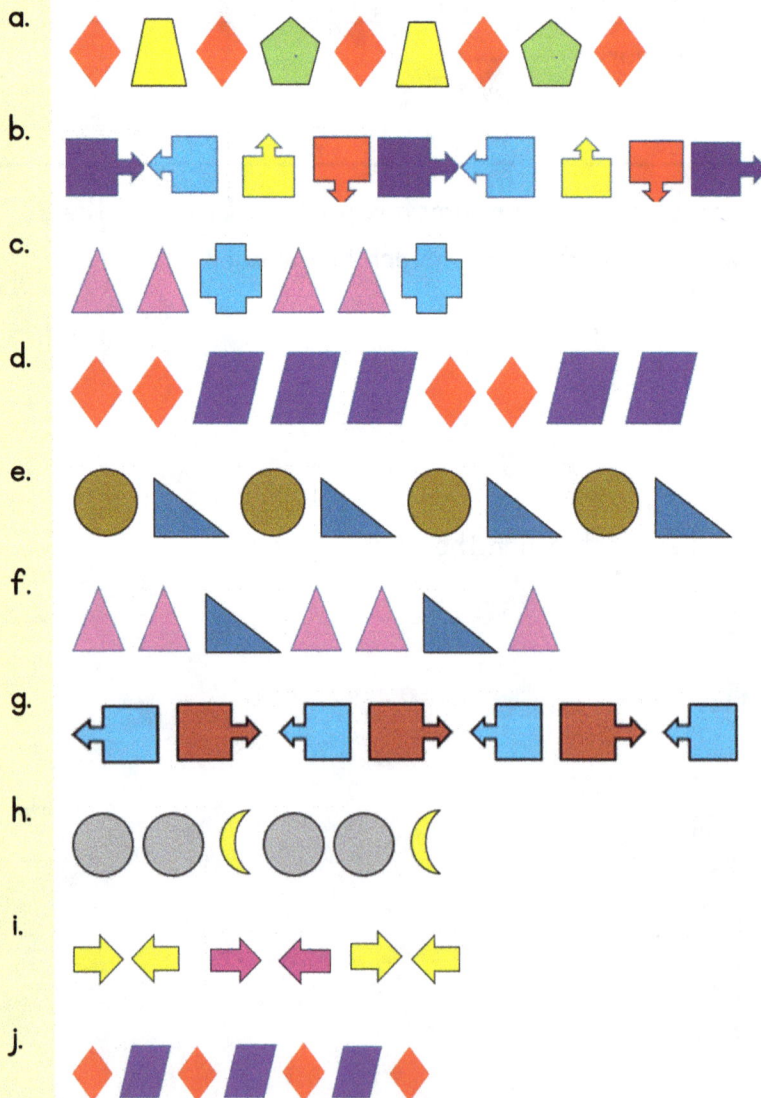

a.

b.

c.

d.

e.

f.

g.

h.

i.

j.

CHEETAH
Connect to Higher Education, Electronic Tools, Aplication and Help

17.3 Associate each term in a pattern with its position in the sequence and express this information in a tabular form.

Let us practice 17.3

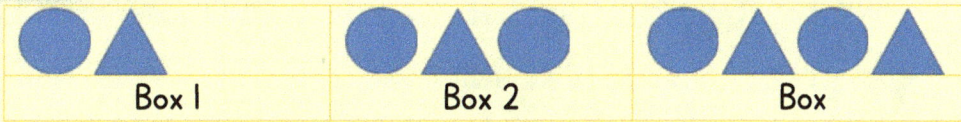

⬤▲	⬤▲⬤	⬤▲⬤▲
Box 1	Box 2	Box

1. A. i. What is the sum of all the shapes in the three boxes? _____

   ii. Draw the set of shapes for box 5 _____

   iii. How many shapes are in boxes 5, 6 and 7 altogether? _____

   iv. How many shapes are in box 8? _____

   v. How many circles are in box 8? _____

   vi. How many triangles are in box 8? _____

B. Use the information from question 1 to complete the table.

Box number	Number of circles	Number of triangles	Total number of shapes
1			
2			
3			
4			
5			
6			
7			
8			

2. This pattern is made by packing boxes neatly together.

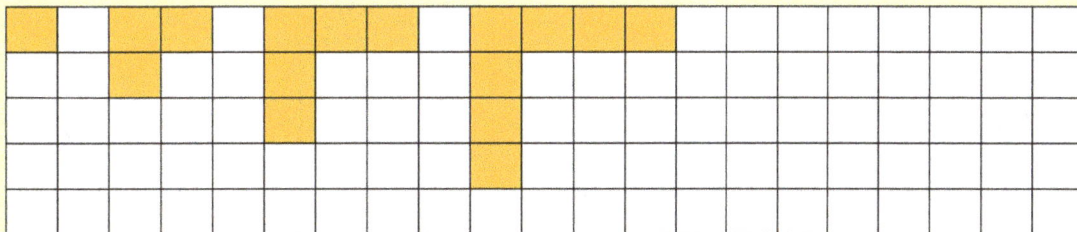

   a. Describe the pattern. _____

   b. How many boxes packed together form each shape on the grid?_____

   c. Draw the 5th shape on the grid.

   d. Guess the number of boxes needed to form shapes 5, 6, 7, 8 and 9. _____

Connect to Higher Education, Electronic Tools, Aplication and Help

3. Complete the following geometric pattern table.

Ordinal numbers	Geometric shapes	Number of red dots	Number of houses	Number of sticks
1st		5		6
2nd		10	2	
3rd				
4th				

Evaluation: Let us see how you did.

Learning outcome	No!	Working on it	Yes!
Did you get all the answers?	😞	😐	🙂
Did you get most questions right?	😞	😐	🙂
Did you retry the question(s) you got wrong?	😞	😐	🙂
Were you able to correct your wrong answers?	😞	😐	🙂
If not, did you seek help from others and/or review the chapter?	😞	😐	🙂

Colour the face that shows how you are doing.

How do I use variables to represent unknown numbers?

Prior learning:

✓ Write number sentences in words.
✓ Use symbols to represent unknown numbers.
✓ Describe number patterns using terms such as "one less" and "one more".

> Before we begin, let's see what you know.

Key vocabulary

Check the words you understand:

☐ coefficient
☐ less
☐ more
☐ number sentence
☐ solution
☐ variables

1. Write in words:

   a. $4 - 2$ _____
   b. $12 + 3 = 15$ _____
   c. 5×6 _____
   d. $20 \div 5$ _____
   e. $3 + 13 - 9$ _____

2. Convert words to numbers:

   a. two more than the number fourteen _____
   b. three less than the number twelve _____
   c. twice the number eight _____
   d. the number eight shared among four people _____
   e. four times a number 'n' added to nine _____

3. Use symbol to represent unknown:

   a. six more than ten _____
   b. eight less than twelve _____
   c. n less than four _____
   d. n greater than four _____
   e. four less than n _____

> If you know these, you should be able to learn what comes next.

18.1. Write algebraic sentences for problems. Express simple sentences and word problems as algebraic expressions.
18.2 Write one- or two-step problems based on information given in a story; then write the correct algebraic sentence and solve the problem.
18.3. Use arithmagons to complete number sentences.

18.1 Write algebraic sentences for problems.

Remember that **number sentences** are written as a combination of numbers and symbols. Examples: $5 + 5 + 10 = 20$
or $\$10 + \$5 + \$1 + \$1 + \$1 + \$1 + \$1 = \$20$
and $9\,m + 4\,m + 5\,m + 2\,m = 20\,m$

Let us practice 18.1

Five students opened their saving boxes. Each had $100 in coins saved. Write 4 different ways in which the coins could have been set out. The first has been done for you.

$10	+	$20	+	$20	+	$50	=	$100	
◯	+	◯	+	◯	+	◯	=	$100	
◯	+	◯	+	◯	+	◯	=	$100	
◯	+	◯	+	◯	+	◯	=	$100	
◯	+	◯	+	◯	+	◯	=	$100	

18.2 Express simple sentences and word problems as algebraic expressions.

An algebraic expression has no equal sign. These expressions can be simplified but not solved. They are made up of **variables** and constants e.g. a, 2b, a2 or 4. The four mathematical operations of +, -, x and ÷ are used in writing algebraic expressions. Here are four things to know about algebraic expressions.

The unknown letter is called a variable. **Example:** a, b or x	The constants are the whole numbers, fractions or decimal numbers. **Example:** 4, ½ or 0.5
The numbers just in front of the variables are called **coefficients** **Example:** the 2 in 2b	Coefficients are numbers that are multiplying the variable. **Example:** 2b is 2 x b

Also, here are some verbal phrases and their algebraic expressions:

Verbal phrases	Algebraic expressions
The sum of 4 and m; 4 more than m; m added to 4; the total of 4 and m	$4 + m$
b takes away two; b minus two; the difference of b and two; two less than b	$b - 2$
The product of m and 3; increase m threefold; the product of m and n	$m \times 3$ or $3m$ $m \times n$ or mn
d divided by 5; d shared among 5; the quotient of d divided by 5	$d \div 5$

Example 1: Zeek bought 4 more June plums than his friend Marlon who bought m June plums. How many June plums did Zeek buy?

Answer: The expression is 4 + b

Example 2: Keisha had d oranges and increased it by 3 folds. She also got 5 more oranges from Kemar. How many oranges did she have in all?

Answer: d x 3 + 5 = 3d + 5. The expression is 3d + 5

Let us practice 18.2

Write an expression for the following.

1. Rohan has five less than his friend Tom who has b. How many does Rohan have? _____

2. Amia has three times as many as her friend who has X. How many does she have? _____

3. Kelvin has half of a number r. What number does he have? _____

4. Alfred has six less than B. How many does Alfred have? _____

5. The amount of money A is $80 more than C. How much is A? _____

6. Tony has 8 more than the number D divided by 2. How many does Tony have? _____

7. Jackie has 8 less than the product of four and a number x. How much does Jackie have? _____

8. Alyssa spent 6 minutes more than Annesha doing a test. If Annesha spent y minutes, how long did Alyssa spend? _____

9. Greg spent 8 minutes less than Damon doing a test. How long did Greg take, if Damon took x minutes? _____

10. Tommy took n minutes and Mary took 2 minutes more than Tommy. How long did they take altogether? _____

18.3 Write one- or two- step problems based on information given in a story; then write the correct algebraic sentence and solve the problem.

An algebraic sentence is a number sentence with one or more unknown numbers written as a letter.

Example 1: Two is added to an unknown and the answer is 8. This problem can be written as the algebraic sentence or algebraic equation   u + 2 = 8.

An equation has two sides: the left-hand side (LHS) and the right-hand side (RHS).

U + 2 =	8
LHS	RHS

So, the equation says LHS = RHS

To keep this balance whatever is done to the left-hand side must be done to the right-hand side.

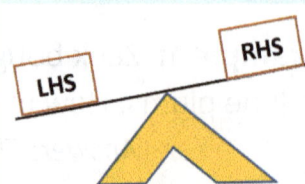

RHS

LHS

To solve an algebraic equation, we have to get the u alone on the left-hand side of the equation. This is done by subtracting the 2 from the LHS. So, the 2 is also subtracted from the RHS as shown.

$$U + (2 - 2) = (8 - 2)$$
$$U + 0 = 6$$
$$U = 6$$

Example 2: I think of a number, then I double the number. My answer is 12. What number did I think of?

Steps:

1. Let the number be letter a.

2. Write the equation as a x 2 = 12 or 2 x a = 12 or 2(a) = 12 or a(2) = 12. So 2a = 12.

3. Remove 2 so the a is by itself to leave the unknown number.

$$a \times 2 \div 2 = 12 \div 2$$
$$2a \div 2 = 12 \div 2$$
$$\frac{\cancel{2}a^1}{\cancel{2}_1} = \frac{\cancel{12}^6}{\cancel{2}_1}$$
$$a = 6$$

I thought of number 6.

Example 3: I think of a number then multiply the number by three then add three to the product. My answer is 9. What number did I think of?

In algebraic sentence this is: x x 3 + 3 = 9

Which gives: 3x + 3 = 6

$$3x + 3 - 3 = 6 - 3$$
$$3x + 0 = 3$$
$$3x = 3$$
$$\frac{\cancel{3}x^1}{\cancel{3}_1} = \frac{\cancel{3}^1}{\cancel{3}_1}$$
$$x = 1$$

Example 4:

Box	-	5	=	12
Box	-	5 + 5	=	12 + 5
		Box	=	17

Next, let us look at some examples of algebraic sentences and how to solve them.

a + 1 km + 1 km = 4 km	x + \$2 + \$6 = \$24
a + 2 km = 4 km	x + \$8 = \$24
a + 2 km – 2 km = 4 km – 2 km	x + \$8 - \$8 = \$24 - \$8
a + 0 km = 2 km	x + \$0 = \$16
a = 2 km	x = \$16

$$a + 5 + 5 + 10 = 28$$

$$a + 20 = 28$$

$$a + 20 - 20 = 28 - 20$$

$$a = 8$$

Pops has two grandchildren. He gave one $30 and the other $20. How much money did he share?

Here the algebraic sentence is $30 + $20 = x.

$$\$30 + \$20 = x$$

$$x = \$30 + \$20$$

$$x = \$50$$

Let us practice 18.3.1

Write algebraic sentences for the following and solve them.

1. There are thirty-five students in the class. The teacher gave each student four tickets to sell. How many tickets were handed out to be sold? _____

2. The teacher gave out 40 pencils and each pencil cost $9. What is the cost for all the pencils given out by the teacher? _____

3. For the class party the teacher collected $100 from each of 40 students for five days. How much money was collected in all for the five days? _____

4. A basket of 35 grapes fell from the table. Only 26 were not damaged. How many were damaged? _____

5. When 84 sweets were divided among six students, how many did each student get?

6. If the class had $85 to buy cupcakes that cost $5 each, how many students gave money to buy a cupcake? _____

7. A large chocolate bar cost $85 and each student contributed $15. How many students contributed, if they collected $5 too much? _____

8. Tickets were sold at $5 each. The money collected for tickets was $65. If 20 tickets were available to be sold, how many were not sold? _____

9. The grade 4A class in Belmont Primary School was expecting to get $9,000 from a cake sale to cover the cost of printing a PEP test for everyone in the class. Unfortunately, they only collected $6,000. If there are 30 students in the class, how much should each student give to cover the shortfall? _____

Let us practice 18.3.2

Write algebraic expressions or an equation and solve it.

1. If 2 more than x gives 8, find x. _____

2. If three more than the product of two and p gives 7, what is the value of p? _____

3. Write an equation for the information: 8 less than the product of 2 and a number is equal to 12. _____

4. Write an expression for this information: A certain number T is divided by three and 9 is added. _____

5. One more than the product of 4 and x gives 9. Find the value of x. _____

6. Four less than the product of three and n gives 5. Find the value of n. _____

CHEETAH
Connect to Higher Education, Electronic Tools, Aplication and Help

7. The product of four and a number x when divided by three gives 15. Write an equation for this information. _____

8. If half of X is 8, what is X? _____

9. If 4 times a number t Is equal to 20, what is t? _____

10. Xavier spent $10 more than 2 times what Mika spent. If Meka spent $20, how much did Xavier spend? _____

Using arithmagons

Arithmagons are polygons with a circle on each vertex and a box number on each side. Each box number is the sum of the two circle numbers on both sides of the box.

A polygon with the numbers at its vertices tells the value on the edges.

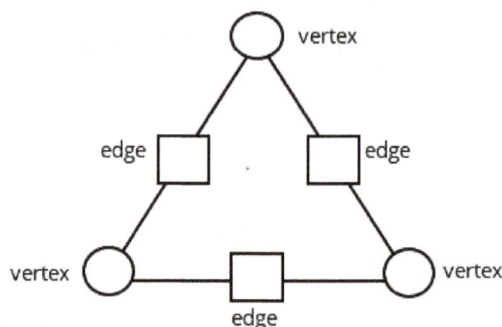

Given the values at the **vertices**, you can work out the numbers at the **edges**.	Given the values at the **vertices and edges**, you can work out the missing numbers.
Example: 1 + 2 = 3; 1 + 3 = 4; 2 + 3 = 5	**Example:** 3 = 1 + 2; 5 = 2 + 3; 3 + 1 = 4

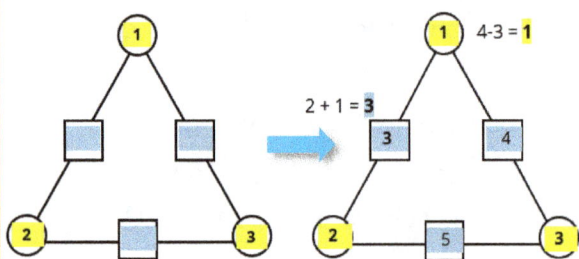

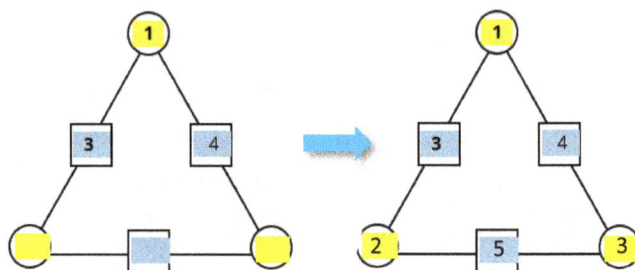

Notice that the numbers in the rectangular boxes are the sums of the numbers in the circles on that side.

Practice adding your own numbers into some of the boxes and circles and solving for the others.

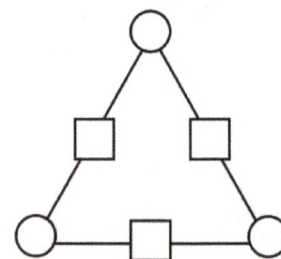

Let us practice 18.3.3

A. Complete the following arithmagons:

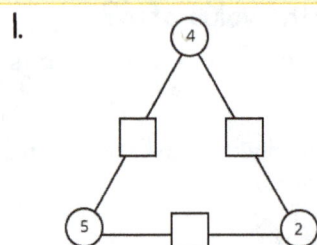

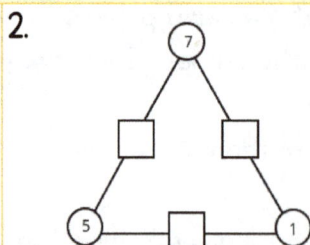

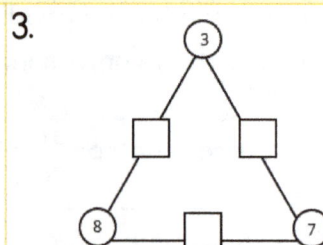

1.

2.

3.

4.

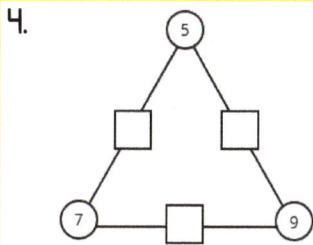

5. Label the vertices and edges as a singular noun.

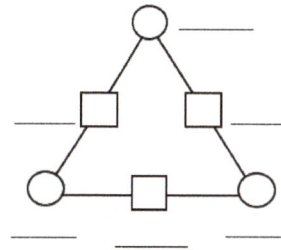

B. Complete the following arithmagons:

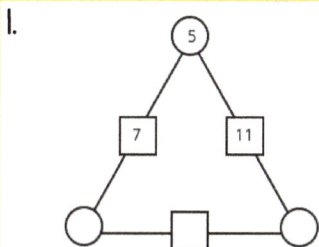

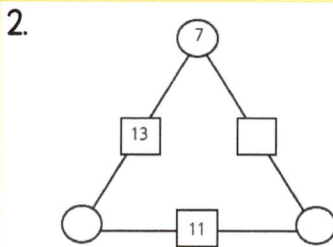

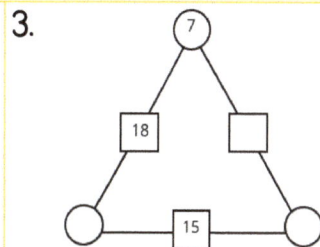

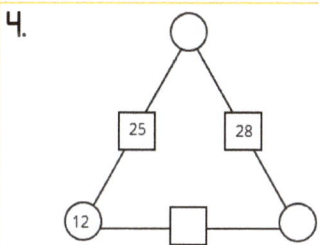

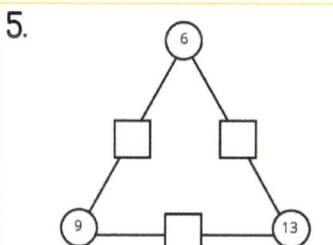

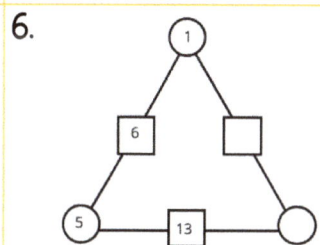

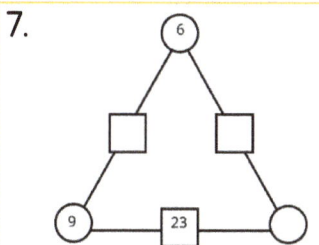

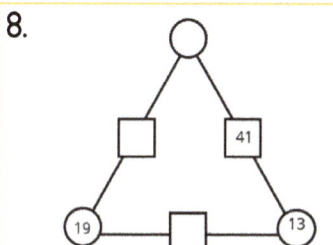

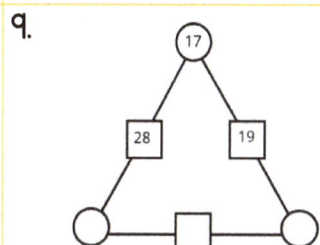

1.

2.

3.

4.

5.

6.

7.

8.

9.

Evaluation: Let us see how you did.

Learning outcome	No!	Working on it	Yes!
Did you get all the answers?	☹	😐	🙂
Did you get most questions right?	☹	😐	🙂
Did you retry the question(s) you got wrong?	☹	😐	🙂
Were you able to correct your wrong answers?	☹	😐	🙂
If not, did you seek help from others and/or review the chapter?	☹	😐	🙂

Colour the face that shows how you are doing.

What do I need to do to find the mean of a set of data?

Before we begin, let's see what you know.

Prior learning:

✓ Interpret a graph.

Key vocabulary

Check the words you understand:

☐ central

☐ tendency

☐ data

☐ information

☐ mean ☐ median

☐ mode

☐ population

☐ sample

☐ statistics

☐ survey

1. Tom is reading a book. The graph shows his progress each day.

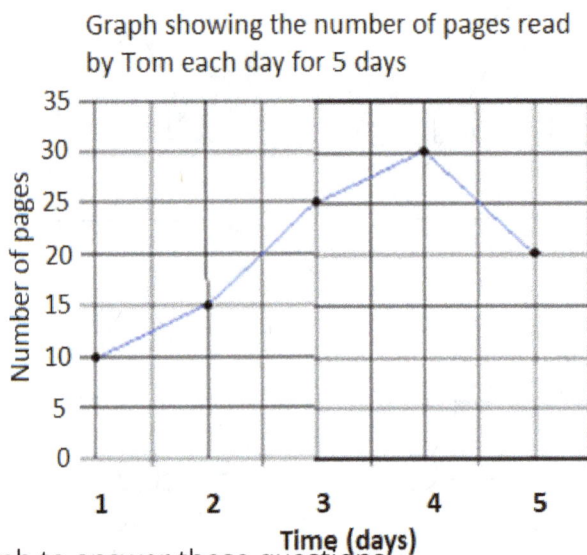

Graph showing the number of pages read by Tom each day for 5 days

Number of pages (y-axis), Time (days) (x-axis)

Use the graph to answer these questions

a. On which day did he read the most pages? _____

b. How many pages did he read in all? _____

c. On which day did he read the least number of pages? _____

2. Jeneil picked oranges every day for 5 days.

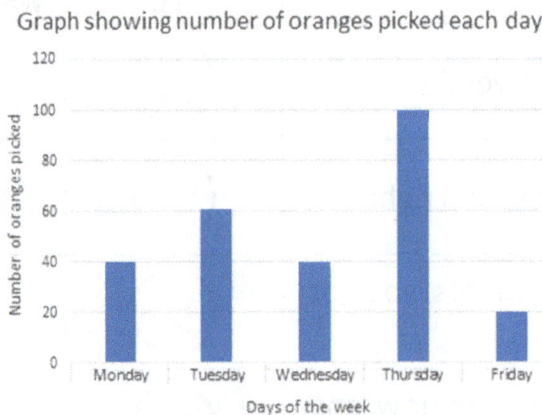

Graph showing number of oranges picked each day

Number of oranges picked (y-axis), Days of the week (x-axis)

If you know these, you should be able to learn what comes next.

a. About how many oranges were picked on Thursday? _____

b. About how many oranges were picked for the week? _____

c. About how many more oranges were picked on Tuesday than on Friday? _____

19.1 Calculate the mean, mode and median of a set of data.
19.2 Calculate the total set given the mean average and the number of addends.
19.3 Solve problems based on the mean.

19.1 Calculate the mean, mode and median of a set of data.

The **mean** or average of a set (group) of numbers is the sum of all the numbers divided by the number of numbers in the set (group).

The **median** is the middle number or the average of the two middle numbers when all the values are arranged in ascending or descending order.

The **mode** of a set of numbers is the number that is seen most often: the most frequent value. A **data** set can have more than one mode.

Example: in a class of 12 students a test was marked out of 10 points. Using the table calculate the mean, median and mode for the students' test results.

Students	Mark
Adrian	3
Brian	8
Camille	4
Diana	7
Everett	7
Francois	10
Gillian	8
Hannah	5
Isabel	8
Jevaughn	9
Kurt	9
Lorenzo	6
Total number of students = _____	**Total of all the marks** = _____

Finding mean, median and mode:

a. Calculate the mean mark for the class test.

$$\text{Mean} = \frac{sum\ of\ all\ student\ grades}{Total\ number\ of\ students}$$

$$\text{Mean} = \frac{3+7+4+7+5+9+8+5+8+8+8+6}{12}$$

$$\text{Mean} = \frac{84}{12}$$

Mean = 7 marks out of 10.

b. Calculate the median mark for the class test.

3, 4, 5, 6, 7, 7, 8, 8, 8, 9, 9, 10

Median = one middle score or $\frac{both\ middle\ scores}{2}$

$$\text{Median} = \frac{7+8}{2} = \frac{15}{2}$$

Median = 7 ½

c. Calculate the mode mark for the class test.

Mode = Most frequent mark

Since 3 students scored 8

Mode = 8

Let us practice 19.1

Find the mean, median and mode of the data given.

	Data	Mean	Median	Mode
i.	2, 4, 4, 5, 7, 8			
ii.	1, 2, 3, 4, 5, 6, 7, 8, 9, 5			
iii.	26, 34, 45, 56, 67, 78			
iv.	12, 13, 13, 13, 14, 15, 15, 18, 22			
v.	18, 22, 23, 25, 32, 35, 40, 44, 58			
vi.	23, 18, 24, 28, 30, 25, 24, 24			
vii.	0, 15, 18, 22, 23, 24, 24, 26, 28			
viii.	48, 56, 62, 73, 73, 89, 96			
ix.	43, 44, 45, 45, 45, 57, 64			
x.	18, 22, 23, 24, 24, 25, 28, 36			

19.2 Calculate the total of set given the mean average and the number of addends.

To find the total of a set of numbers we multiply the mean average by the number of addends.

Example: Given the mean average of a set is 4 and the number of addends is 6, calculate the total of all the numbers in the set.

$$Total = 4 \times 6 = 24$$

Let us practice 19.2

Calculate the total of the following sets of numbers given the mean average and the number of addends.

	Mean average	Number of addends	Total
i.	18	3	
ii.	8	4	
iii.	8	7	
iv.	12	5	
v.	15	8	
vi.	12	2	
vii.	16	5	
viii.	32	4	
ix.	15	5	
x.	25	4	

19.3 Solve problems based on the mean.

Sometimes when you have a set of numbers, one is missing from the set. To find the missing number, we first calculate the total of all the numbers then subtract the sum of the known numbers.

Example: Find the missing number in this data set of 4 numbers, 12, 10, _, 8, if the mean of the set is 9.

How to find the missing number?

a. Find the sum of all the numbers in the data set. Sum = 9 x 4 = 36

b. Find the sum of the given numbers. Sum = 12 + 10 + 8 = 30

c. Find the missing number. Missing number = 36 – 30 = 6

Missing number = total of all the numbers - the sum of the known numbers

Let us practice 19.3.1

1. Solve the following questions.

a. Find the missing number in a set of four numbers. Three of the numbers are 2, 4 and 8, and the average is 5.	
b. The mean of four numbers is 9 and three of the numbers are 8, 12 and 10. Find the 4th number.	
c. The average of five numbers is 8. Find the missing number in the set 12, 9, 5, x, 13.	
d. The average of four numbers is 8. If three of the numbers are 9, 11 and 7, what is the 4th number?	
e. A set of four numbers has an average of 25. If three of the numbers are 14, 36, 12, find the 4th number.	
f. Forty (40) is the average of five numbers. If four of the numbers are 26, 68, 57 and 18, what is the 5th number?	
g. The average of nine numbers is 12. If eight of the numbers are 4, 9, 11, 17, 21, 14, 3 and 13, what is the 9th number?	
h. The average of three numbers is 38. Two of the numbers are 58 and 47. What is the third number?	
i. Over four working days, Tajay worked 10 hours on Monday, 8 hours on Tuesday and 12 hours on Wednesday. If the average hours worked was 10 hours, how many hours did he work on Thursday?	
j. Michael bought 5 sweets at a shop. The prices of the first four sweets were $30, $25, $10 and $15. If the average sweet cost $20, how much did he pay for the 5th sweet?	

2. The chart shows the number of sweets bought by children at a tuck shop each day. Each block is one sweet.

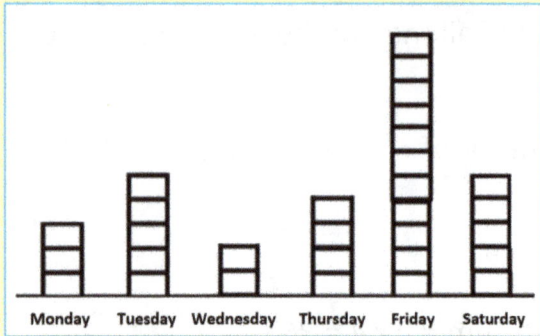

a. How many sweets were bought on Thursday? _____

b. On what day was the least number of sweets bought? _____

c. What is the mean number of sweets bought for the days?_____

d. What is the median number of sweets bought for the days? _____

e. What is the modal number of sweets bought for the days? _____

3. The chart shows the number of boxes of fruits sold each week. Each block is one box of fruits.

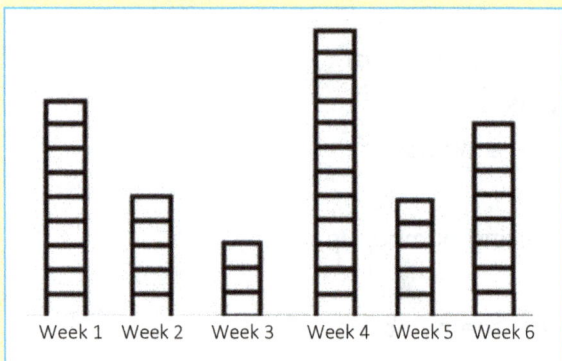

a. On what week is the most boxes of fruits sold? _____

b. What is the total number of boxes of fruits sold over the 6 weeks? _____

c. What is the mean number of boxes of fruits bought for the weeks? _____

d. What is the median number of boxes of fruits bought for the week? _____

4. The chart shows the number of overtime hours a cashier worked at a supermarket for the week. Each block is one overtime hour.

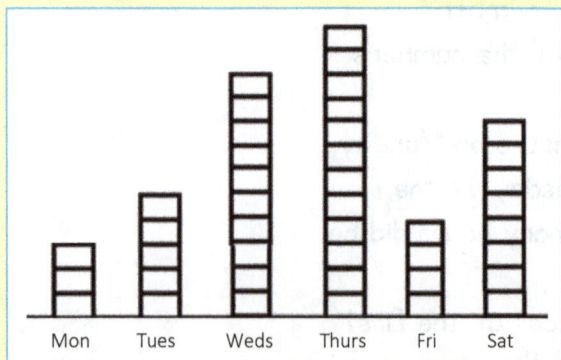

a. On what day of the week was the least overtime hours worked? _____

b. What is the total number of overtime hours worked for the week? _____

c. What is the mean number of overtime hours worked for the week? _____

d. What is the median number of overtime hours worked for the week?

5. A survey was done on 12 teachers from a primary school to see what type of movie they preferred to watch during their recreation time. Each block is one teacher's vote.

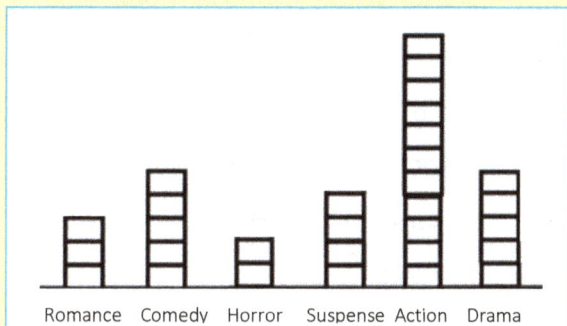

Romance Comedy Horror Suspense Action Drama

a. What type of movie was the favourite among the teachers? _____

b. What type of movie was the least favourite among the teachers?_____

c. What is the mean number of teachers who voted for a movie? _____

d. What is the median number of teachers who chose at least one of the movies shown? _____

e. What is the mode for the survey on the teachers' movie preferences? _____

6. Students in grade 4 were asked to choose their favourite type of cookie for a bake sale. Each block is the response from two students.

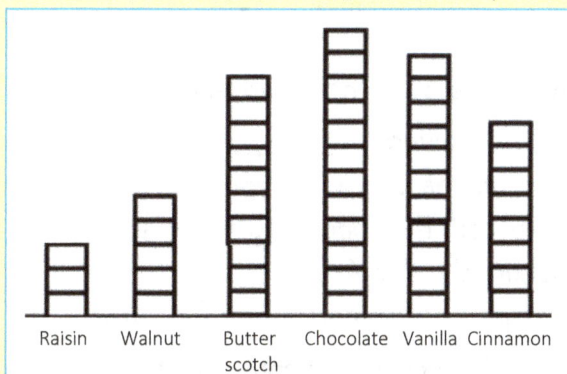

Raisin Walnut Butter scotch Chocolate Vanilla Cinnamon

a. What type of cookie was the second favourite among the students? _____

b. Based on the chart, do you think most grade 4 students like raisin? _Yes / No_

c. What is the total number of students who chose a cookie? _____

d. Based on the chart, which cookies would you recommend be sold at the bake sale and why? _____

e. How can you use a survey like this to help your school raise money? _____

7. This survey collected information on the recreational activities' tourists preferred. This information collected is represented in the chart. Each block is the response from two tourists.

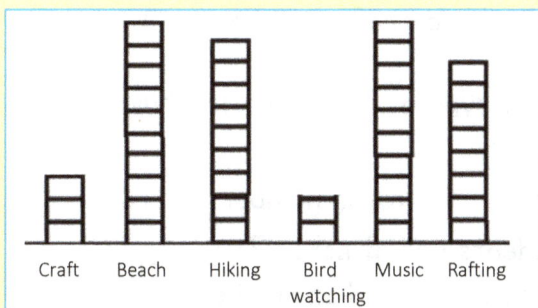

Craft Beach Hiking Bird watching Music Rafting

a. What type of recreational activity was the favourite among the tourists? _____

b. Based on the chart, do you think most tourists visit for bird watching? _Yes / No_

c. What is the mean number of tourists who chose at least one of the activities? _____

d. What is the median number of tourists who chose at least one of the activities?

e. Based on the chart, which four activities would you promote on our new visitors' calendar and why?

8. The chart shows the breakfast foods served at a canteen. Each block is 10 purchases.

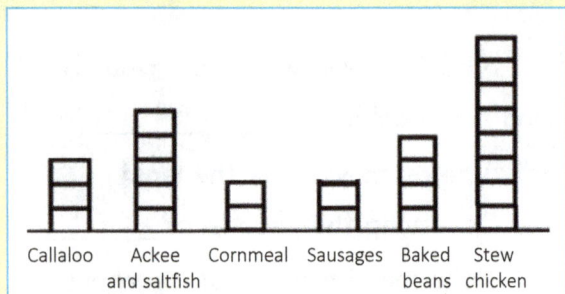

a. What type of breakfast food was the most sold? _____

b. Based on the chart, how many purchases of cornmeal porridge were made? _____

c. What is the mean number of breakfast foods purchased? _____

d. What is the median number of breakfast foods purchased? _____

e. How many more people preferred stew chicken over calaloo? _____

9. In a survey done in a library, data was collected on the types of printed materials borrowed. This data is represented in the chart. Each block is 3 printed materials borrowed.

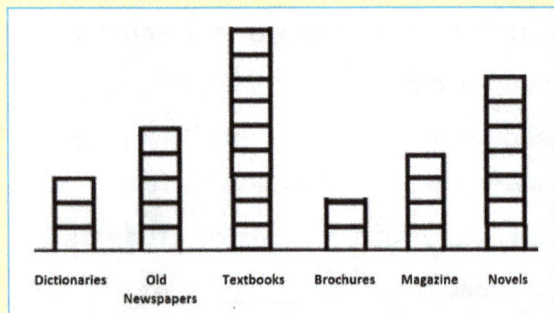

a. What type of printed material was the most borrowed? _____

b. Based on the chart, do you think most people visit the library to borrow brochures? _Yes / _No

c. What is the mean number of printed materials borrowed from the library? ___

d. What is the median number of printed materials borrowed from the library? ___

e. The library will be featured on a community billboard. Based on the chart, which three activities would you recommend be promoted and why? _____

10. In a survey, data was collected on the preferred fruits to eat at school. This data is represented in the chart. Each block is 1 vote for a preferred fruit.

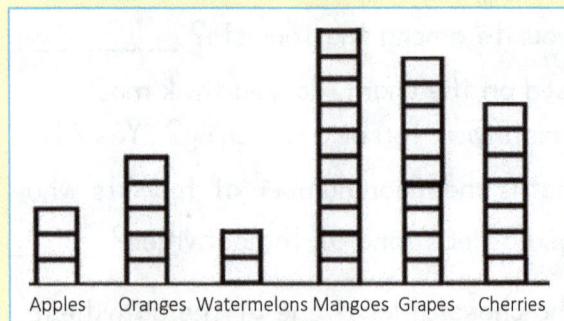

a. What type of fruit was the favourite? _____

b. What is the mean number of fruits students eat at school? _____

c. What is the median number of fruits students eat at school? _____

d. Based on the chart, what are the three least likely fruits to purchase for the students, if you want them to eat fruits?

11. In a survey, data was collected on clubs students preferred to join at school. This data is represented in the chart. Each block is 10 votes for a preferred club.

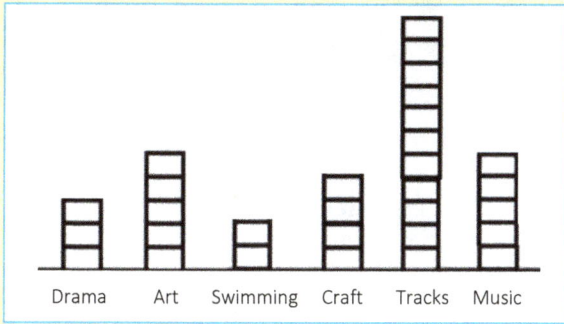

Drama Art Swimming Craft Tracks Music

a. What club was the least favourite among the students? _____

b. Based on the chart, do you think most students like drama club? _Yes / No_

c. What is the mean number of students who are voting? _____

d. What is the median number of students? _____

e. Based on the chart, which club is the most popular and how many students are in that club? _____

12. Maria's marks on 8 spelling tests are shown in the diagram.

a. Which number is Maria's most common score? _____

b. Would it be correct to say that Maria usually scores about 7 in her spelling test? _____

c. Which of the following do you think would be Maria's most likely score in the 9th test?   a. 9   b. 8   c. 7   d. 6

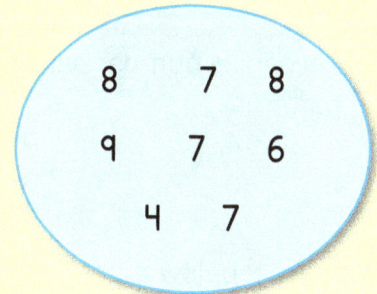

8   7   8
9   7   6
4   7

13. Diagram shows the number of people attending a prayer meeting at church over 8 days.

a. Which number best represents the number of people who attend each day? _____

b. Would it be correct to say that 6 people usually attend the meeting? Explain why: _____

c. Which of the following do you think would be the number of people attending the meeting held on the 9th day? a. 3   b. 4 c. 5   d. 6

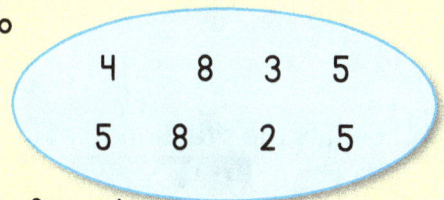

4   8   3   5
5   8   2   5

Evaluation: Let us see how you did.

Learning outcome	No!	Working on it	Yes!
Did you get all the answers?	☹	😐	☺
Did you get most questions right?	☹	😐	☺
Did you retry the question(s) you got wrong?	☹	😐	☺
Were you able to correct your wrong answers?	☹	😐	☺
If not, did you seek help from others and/or review the chapter?	☹	😐	☺

Colour the face that shows how you are doing.

How do I collect, organise, display and interpret information?

Prior learning:
- ✓ Collect and record data.
- ✓ Recognize number patterns.
- ✓ Differentiate between sample and population.

Before we begin, let's see what you know.

Key vocabulary

Check the words you understand:

- ☐ data ☐ graph
- ☐ information
- ☐ interpret
- ☐ interviewing
- ☐ observation
- ☐ population
- ☐ presentation
- ☐ questionnaire
- ☐ sample
- ☐ sampling
- ☐ techniques
- ☐ survey

1. Complete the table.

Transport	Tally	Frequency
Walk	JHT JHT JHT	
Bus		12
Car	JHT II	
Bike		6

2. Recognize number patterns. Which number comes next in the patterns?
   a. 20, 14, 8 …
   b. 2, 9, 16, …
   c. 5, 10, 15…
   d. 80, 40, 20 …
   e. 2, 4, 8 ….

3. Write R below the instrument used to record data and C below the instrument on/in which data is collected.

Pencil	Paper	Tally sheet	Questionnaire	Pen	Graph paper	Charts	Survey	Tables

4. Identify the situations that involve data collection by ticking the images.

☐   ☐   ☐   ☐   ☐   ☐

5. Which of these pictures, P or S correctly shows a sample (S) and a **population** (P)? Circle X or Y.

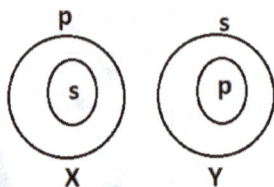

P — S (X)   S — P (Y)

If you know these, you should be able to learn what comes next.

6. A population of 12 people and five **samples** [(a) to (e)] are shown. Select the best sample for the population. Explain why that sample was chosen.

a.   b.   c.

d.   e.

_____

_____

🎯 20.1 Read and interpret bar graphs, line graphs, picture graphs and pie charts.
20.2 Present data using pictographs and bar graphs.
20.3 Convert a pictograph into bar graph or vice versa

20.1 Read and interpret bar graphs, line graphs, picture graphs and pie charts.

When **data** is collected, it can be recorded as a bar chart, line **graph**, pictograph (picture graph) or pie chart. These **techniques** allow data to be easily **interpreted**.

Bar graph (chart) example
Collect the data red = 50, blue = 70, grey = 20, yellow = 40. Now use the data to answer the question.

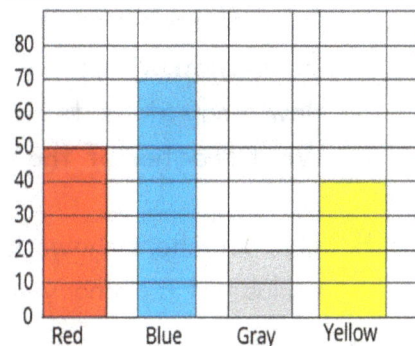
Bar chart showing the favourite colours of students in Grade 4

a. Which colour did most students prefer? *blue*
b. Which colour did the least number of students prefer? *grey*
c. How many more students chose blue than grey?
   70 – 20 = 50 students
d. By how many times was yellow greater than grey?
   40 ÷ 20 = 2 times
e. How many students are in grade 4, if 4 students did not choose a colour? *50 + 70 + 20 + 40 + 4 = 184 students*

Let us practice 20.1.1

1. This bar chart shows the number of pets children in a grade four class have at home. All 40 students were asked what pets they had.

   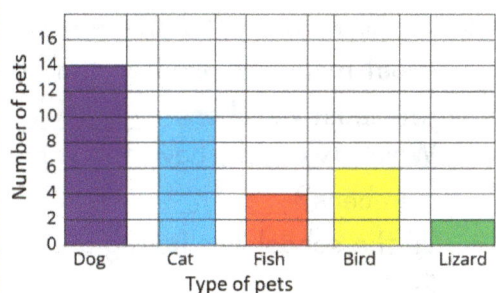

   a. What is the total number of pets for the class? _____
   b. Which pet was the most popular? _____
   c. By how many times was dog a more popular pet than lizard? _____
   d. What fraction of the student population had a cat? _____
   e. How many students did not have a pet? _____

2. This bar chart shows the number of computers sold at each of five shops at a technology fair.

   Shops at a technology fair

   a. Which shop sold the most computers? _____
   b. Which shop sold the least computers? _____
   c. By how many computer sales was the most greater than the least? _____
   d. How many computers were sold in all? _____
   e. Which two shops sold the same number of computers? _____

3. This bar chart shows the number of students in a class who were born in each month.

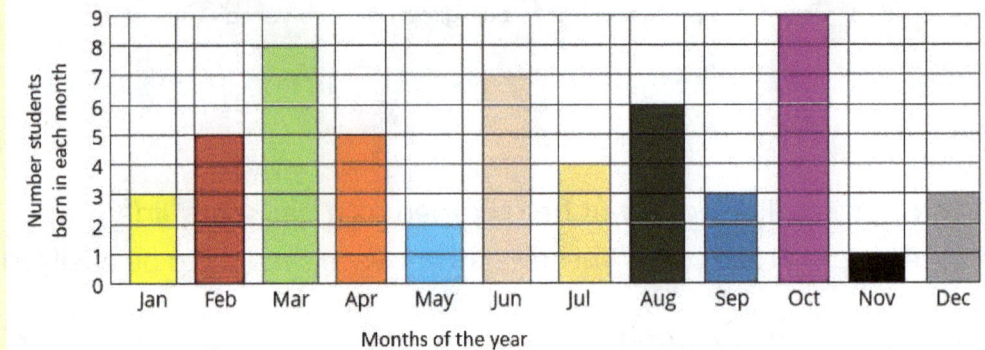

a. What is the total number of students in the class? _____
b. In which month do they have to sing Happy Birthday the most times? _____
c. In which month do they have to sing Happy Birthday the least times? _____
d. How many students had a birthday while school was out for summer holidays? _
e. What fraction of the student population was born in March? _____

4. This bar chart shows the number of students in a grade 4 class who participate in clubs at a primary school.

a. If all the students in the class participate in at least one club, how many students are in the class? _____
b. Which club has the most members from that class? _____
c. By how many students was the club with most members greater than the club with the least members? _____
d. Which two clubs have the same number of members? _____

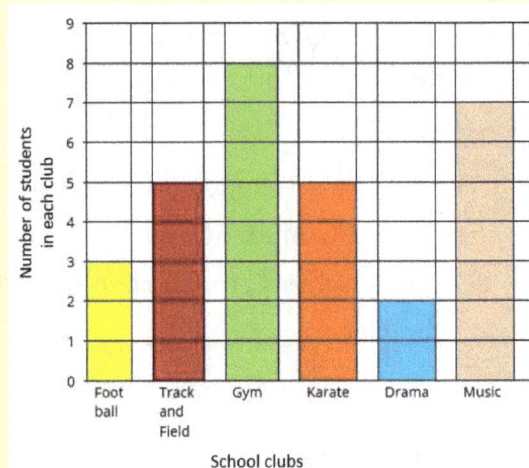

e. If the school was having a concert, which two clubs would be most active? Explain: _____

5. The bar chart below shows the number of animals on a farm.

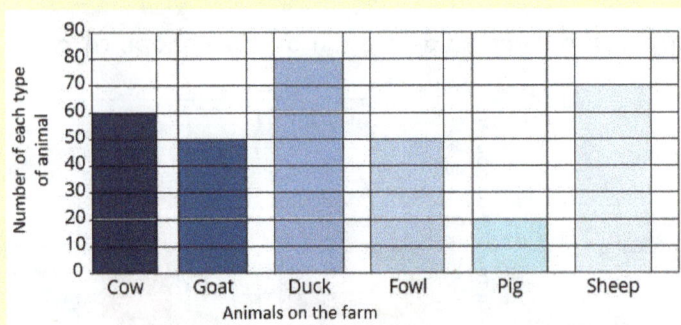

a. How many animals are on the farm? _____
b. Which type of animal is the most abundant on the farm? _____
c. Which type of animal is the least abundant on the farm? _____

d. How many more ducks are on the farm compared to pigs? _____
e. If the farmer sold ½ of the animals on the farm, how many animals would she have left? _____

Line graph example

Line graph showing the number of pages Dwayne read from a book each day over 5 days

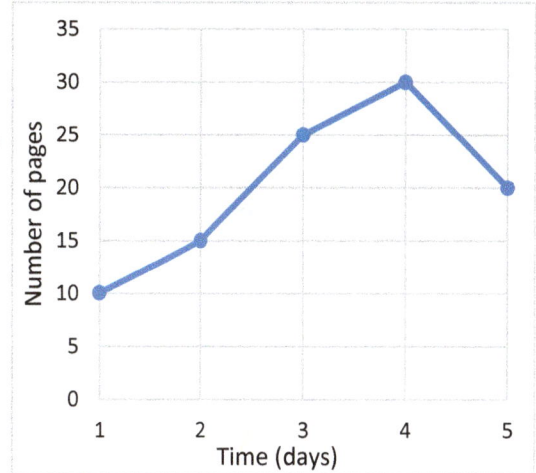

a. How many pages did he read over 5 days? *100 pages*
b. On which day did he read the most pages? *Day 4*
c. On which two days did he read the least number of pages? *Day 1 and 2*
d. By how many times was the number of pages read on the 4th day more than the number of pages read on the 1st day? *40 ÷ 10 = 4 times*
e. If he has 60 more pages to read, how many pages in all are in the book? *10 + 15 + 25 + 30 + 20 + 60 = 160 pages*

Let us practice 20.1.2

1. The line graph shows the scores of 10 students (S1 to S10) on a mathematics test.

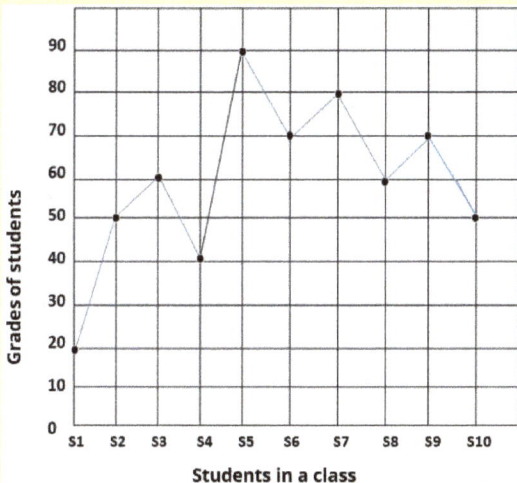

   a. What was the highest score on the test? ____
   b. Which student got the highest score? _____
   c. What was the lowest score on the test? ____
   d. Which students got the same score on the test and how much did they get? _____
   e. What was the mean average score for the 10 students? _____

2. Line graph below shows the number of books sold at a local bookstore each day in a week.

   a. What was the largest number of books sold in the week? _____
   b. What day was the least number of books sold? _____
   c. On which days were the same number of books sold? _____
   d. What is the difference between the number of books sold on Friday and the number of books sold on Thursday? _____
   e. What was the mean average number of books sold in the last 2 days of the week? _____

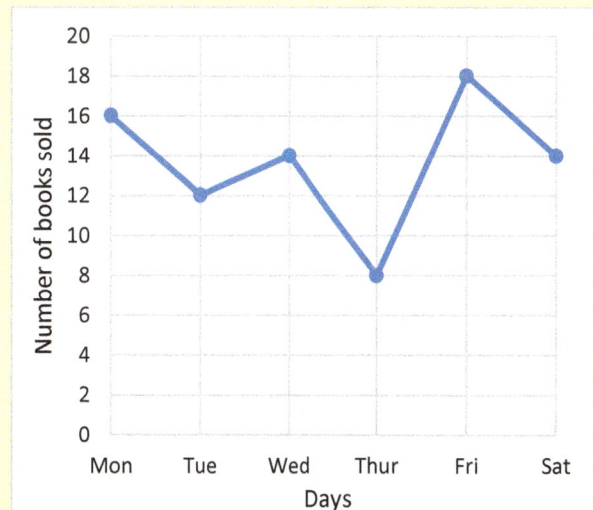

3. The number of goals scored in each of six football matches is shown on a line graph.

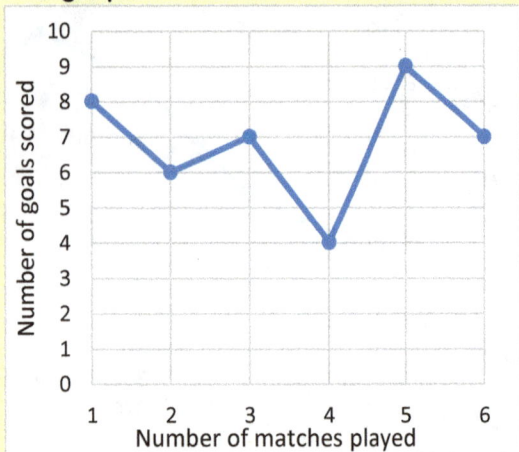

a. What was the highest number of goals scored and in what match? _____

b. What was the lowest number of goals scored and in what match? _____

c. State the total number of goals scored in all six matches. _____

d. In which matches were the same number of goals scored? _____

e. What was the mean average score for the first three matches played? _____

4. In a car mart, the data on the number of cars sold each week was recorded to know if the number of sales was increasing, the same or going up and down each week.

a. What was the number of cars sold in week 2? _____

b. During which week was the least number of cars sold? _____

c. What is the total number of cars sold in all six weeks? _____

d. Use your finger to trace the movement of the line. Was the number of sales increasing, staying the same or going up and down each week? _____

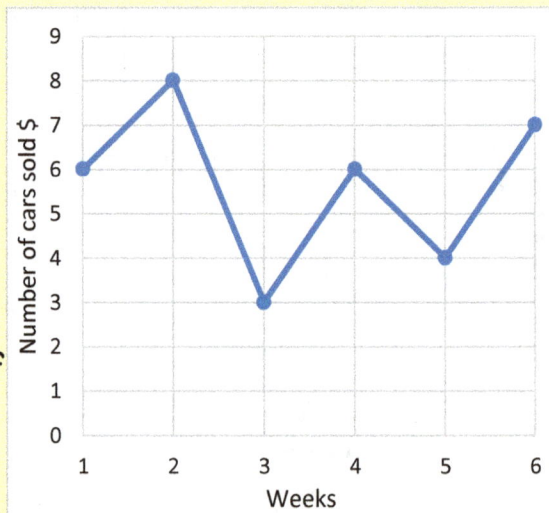

e. During which two weeks was the difference in cars sold equal to one? _____

5. A student was asked to plant a tree from a seed and record the number of leaves which appear during the first nine days after planting. The line graph shows the data recorded.

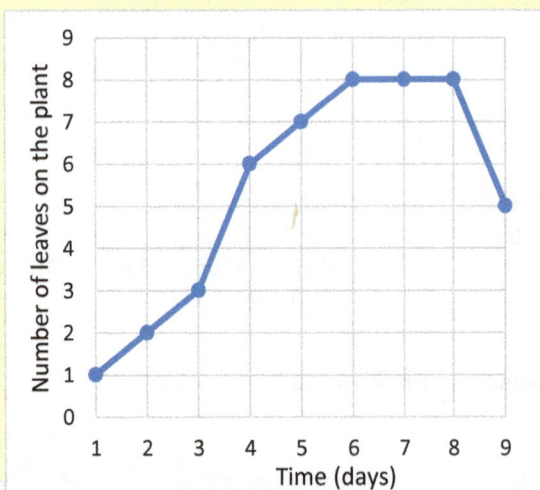

a. How many leaves were seen on day 7? _____

b. Between which two days did the plant grow three leaves? _____

c. Between which days were there no new leaves seen? _____

d. What happened on the 8th day? _____

e. What was the number of leaves on the tree on the 9th day? _____

Pictograph example

Monday = 25, Tuesday = 10, Wednesday = 15
Thursday = 10, Friday = 40

a. How many people were vaccinated on Monday?
   25 people
b. How many people were vaccinated for the week?
   25 + 10 + 15 + 10 + 40 = 100
c. On what days were the same number of people vaccinated? Tuesday and Thursday
d. On what day was the most people vaccinated?
   Friday
e. How many more people were vaccinated on Friday, than on Wednesday? 40 – 15 = 25

Pictograph showing number of people vaccinated for one week at a clinic

Each 👤 represents 5 people.

Days of the week	Number of people vaccinated
Monday	👤👤👤👤👤
Tuesday	👤👤
Wednesday	👤👤👤
Thursday	👤👤
Friday	👤👤👤👤👤👤👤👤

Let us practice 20.1.3

1. In a grade 4 class, students were asked to go to the following locations and record the number of grasshoppers seen in each location.

Pictograph of grasshoppers found at different locations

Locations	Number of grasshoppers found
School	🦗🦗🦗🦗
Church	🦗
Home	🦗🦗🦗
Neighbour's yard	🦗🦗🦗🦗🦗🦗
Park	🦗🦗🦗
Key: Each 🦗 = 4 grasshoppers	

a. How many grasshoppers were found at school? _____
b. What is the difference between the number of grasshoppers found in the neighbour's yard and at church? _____
c. What is the difference between the number of grasshoppers found at the park and the church? _____
d. What was the total number of grasshoppers observed? _____
e. Where would you most likely to find birds? __

2. A local farmer brought a basket of 41 apples for a grade 4 class. Five boys in the class were asked to share the apples among their classmates.

Pictograph showing the number of apples the boys kept for themselves from the basket of apples

Boys	Number of apples taken
Dwayne	🍎🍎🍎🍎
Devin	🍎🍎🍎
Desmond	🍎
Daniel	🍎🍎🍎
Dave	🍎🍎🍎🍎
Key: Each 🍎 = 2 apples	

a. How many apples did Devin take? _____
b. Who took the most apples and how many did he take? _____
c. Who took the least number of apples and how many apples did he take? _____
d. What was the total number of apples taken by the five boys? _____
e. If the remaining apples were shared with other students in the class, how many apples were given to the other students in the class? _____

3. Five local shoe stores entered into a competition to see which could sell the most shoes.

Pictograph of shoes sold

Shoe Stores	Number of shoes
Happy Feet	👟👟👟
The Shoe Shop	👟👟👟👟
Discount Centre	👟👟👟👟👟👟
Everything Shoe	👟👟👟
Walk in Style	👟👟

Key: Each 👟 = 10 pairs of shoes

a. How many pairs of shoes were sold at the Discount Centre? _____
b. Which shoe store sold the least number of shoes? _____
c. What was the total number of shoes sold? _____
d. What was the average number of shoes sold in the competition? _____
e. Which shoe store won the competition and why? _____

4. Each child in a grade 4 class was asked to watch one movie and write a book report describing the movie they watched.

Pictograph of movies grade 4 children watched

Movie watched	Number of children
Happy Feet	🧍🧍🧍
Little Mermaid	🧍🧍🧍🧍🧍🧍🧍
Dragon's Tales	🧍🧍🧍
Despicables	🧍🧍🧍🧍🧍🧍
Little Critters	🧍

Key: Each 🧍 = 2 children

a. Which movie did most children watch? ___
b. How many children watched Little Critters? _____
c. How many children watched Dragon's Tale? _____
d. If you were to plan a movie day, which movie are you least likely to show and why? _____
e. If all the children in the grade 4 class gave in a book report, how many book reports would the teacher receive? ____

5. Five parking lots were observed. The number of cars in each parking lot was used to create the pictograph shown.

Car parked in five parking lots

Parking lots	Number of cars
A	🚗🚗
B	🚗🚗🚗🚗🚗🚗
C	🚗🚗🚗
D	🚗🚗🚗
E	🚗🚗

Key: Each 🚗 = 4 cars

a. Which parking lot had the most cars? ___
b. How many cars were in parking lots? ____
c. How many cars were in parking lot E? ___
d. Which parking lot has 10 cars? _____
e. What is the mean average number of cars parked in the 5 parking lots? _____

Pie chart example

a. What was the most common means of transport? car
b. By how much was the number of students traveling by car greater than the number of students traveling by bus? 2 times
c. What fraction of students traveled to school in a car? $\frac{1}{2}$
d. What fraction of students traveled to school in a bus? $\frac{1}{8}$
e. If 16 students traveled to school in a car, how many walked to school? 4

Pie chart of survey data showing how children in a grade 4 class travelled to school.

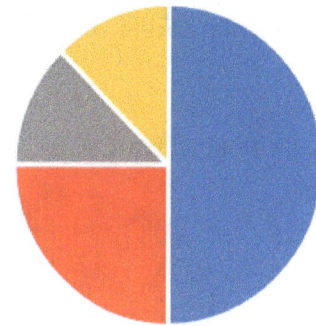

■ Car ■ Bus ■ Bicycle ■ Foot

1. A grade 4 class plans to sell cups of cornflakes and milk to raise funds for a class trip. To choose what types of milk to buy, each student picked their favourite type of milk in a survey. The Pie chart shows the results of survey.

Favourite types of milk chosen by students in a class

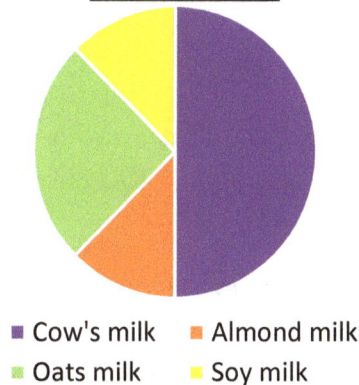

■ Cow's milk   ■ Almond milk
■ Oats milk   ■ Soy milk

a. Which was the most popular type of milk? ____
b. What fraction of students choose almond milk? _____
c. Which two types of milk were equally favoured by the students? _____
d. How many students choose oats milk as their favourite type of milk if there are 40 students in the class? _____
e. How many students did not choose a favourite type of milk? _____

2. Every year at Middleset Primary School a different teacher is chosen to lead the drama club. The survey shows the results of a survey on the last four teachers to lead the drama club.

Pie chart of favourite drama club lead teachers

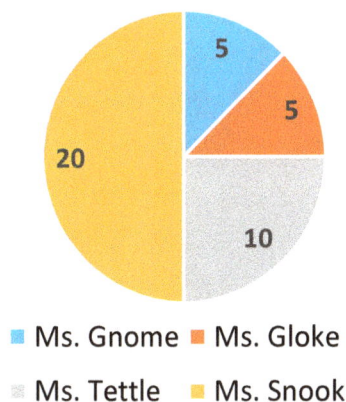

■ Ms. Gnome ■ Ms. Gloke
■ Ms. Tettle ■ Ms. Snook

a. How many students preferred Ms. Snook's drama club? _____
b. Which teachers were the least favourite drama club leaders? _____
c. What fraction of students choose Ms Tettle's drama club as their favourite? _____
d. By how much was the favourite teacher's votes greater than the least favourite teacher's vote? _____
e. Based on these results, which teacher's drama club would have the most children participating in it? Explain why: _____

3. A school was planning a movie day for all the children. The principal wanted to know which type of movie children would most enjoy. The pie chart shows the results of a survey done on 100 primary school children on most disliked types of movies.

Pie chart of types of movies students disliked

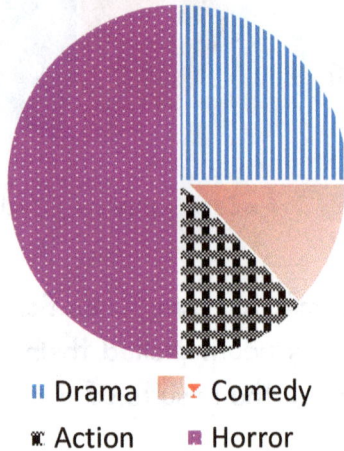

Drama   Comedy
Action   Horror

a. What was the most disliked type of movie? _____

b. How many students disliked drama? _____

c. What fraction of students did not choose action movies as their most disliked type of movie? _____

d. What were the two least disliked type of movies? _____

e. Which types of movie would you recommend the principal choose for movie day? _____

4. This pie chart shows the results of a survey on the number of children who could correctly identify five traditional Jamaican snacks from a group of 100 children.

Number of children who could identify all five traditional Jamaican snacks

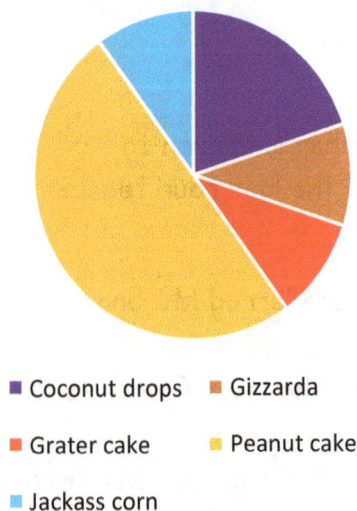

Coconut drops   Gizzarda
Grater cake   Peanut cake
Jackass corn

a. What fraction of the children could correctly identify peanut cake? _____

b. How many children does this fraction represent? _____

c. How many children did not identify peanut cake correctly? _____

d. By how much was the coconut drops correctly identified over the gizzarda? _____

e. Complete the table showing the fraction and number of children who correctly identified each traditional snack.

Traditional snack	Fraction	Number of children
Coconut drops		
Gizzarda		
Grater cake		
Peanut cake		
Jackass corn		

5. Pie chart showing how Janeil's mother spent her salary of $100 for the week.

a. What does Janeil's mother spend most of her salary on? _____

b. Of the remainder, how much money is spent on food? _____

c. Janiel's mother spends twice as much on food as on the children. Estimate what fraction of Janiel's mother's salary is used to pay for food? _____

d. Complete the table showing the fraction and money spent on each expense.

Expenses	Fraction	Number of children
Rent and bills	$\frac{5}{10}$ or $\frac{1}{2}$	
Food		$20
Medical expenses		
Children		
Savings	$\frac{1}{10}$	$10

Legend:
- Rent and bills
- Food
- Medical
- Children
- Savings

e. Does Janeil's mother have any money left over for herself? Explain why: _____

20.2 Present data using pictographs and bar graphs

Data can be taken from a pictograph to create a data table then used to create a bar graph as shown. A bar chart can also be used to create the data table and a pictograph. Bar graphs are drawn on a grid, while pie charts are drawn on plain paper.

Example: Sports day was coming up and the canteen manager wanted to buy 1000 patties for the students. To know how much of each patty to buy, the canteen workers did a survey with 100 students on which type of patty they preferred. The pie chart on the left was presented to the manager. From this, the manager could see that about 5/10 (½) of the students would want beef patties, 2/10 would want chicken and 3/10 would be shared among soy, vegetable and shrimp patties.

Pie chart

Student votes for each type of patty

Legend: Beef, Chicken, Soy, Vegetable, Shrimp

Data table

Patties	Fraction	Votes
Beef	5/10	50
Chicken	2/10	20
Soy	1/10	10
Vegetable	1/10	10
Shrimp	1/10	10

Bar graph

Number of student votes

When each fraction is multiplied by 100 (number of children in survey), the number of votes for each type of patty is obtained. The bar chart can then be created from the number of votes for each type of patty.

Let us practice 20.2

Create a bar chart from the following pie chart.

1. Pie chart of favourite types of milk chosen by 40 students in a class.

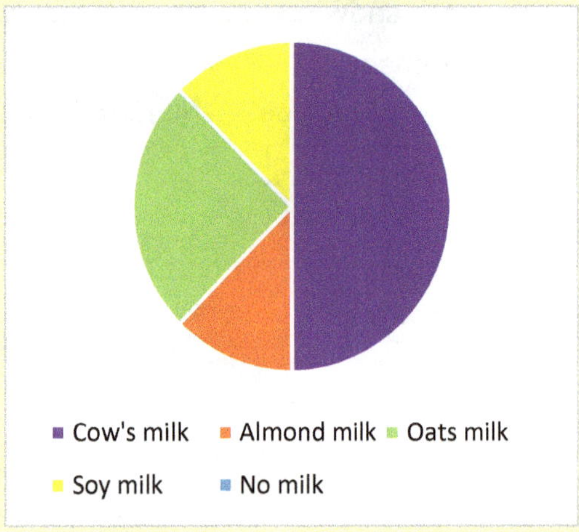

Bar Chart showing _____

- Cow's milk ■ Almond milk ■ Oats milk
- Soy milk ■ No milk

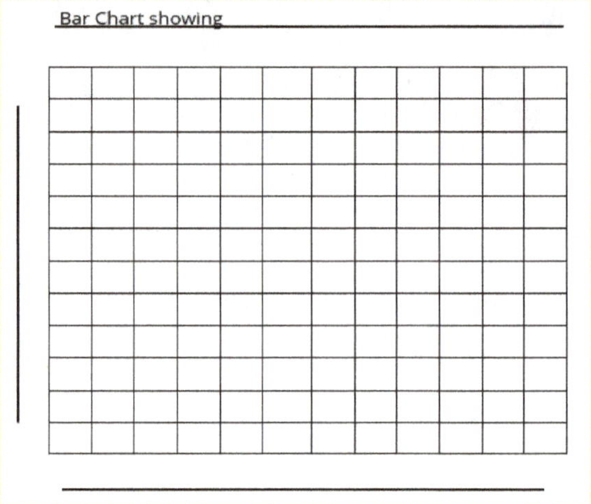

2. Pie chart showing the number of student votes for favourite drama club teachers.

Bar Chart showing _____

- Ms. Gnome ■ Ms. Gloke
- Ms. Tettle ■ Ms. Snook

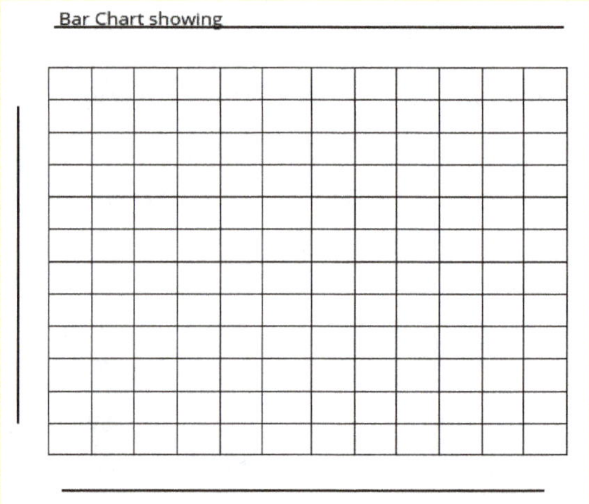

3. Pie chart of types of movies 200 students disliked.

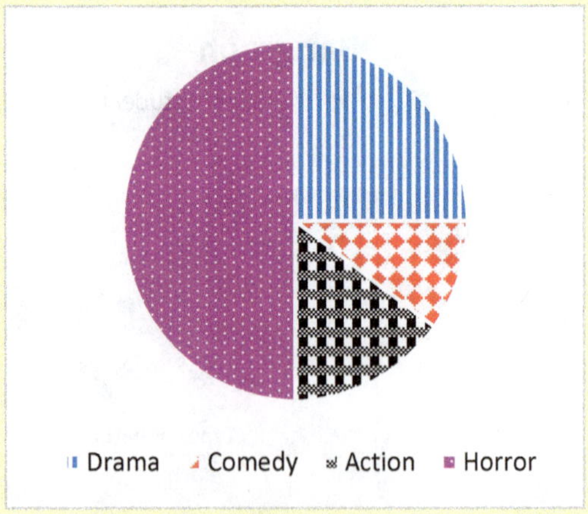

Bar Chart showing _____

■ Drama ◢ Comedy ◣ Action ■ Horror

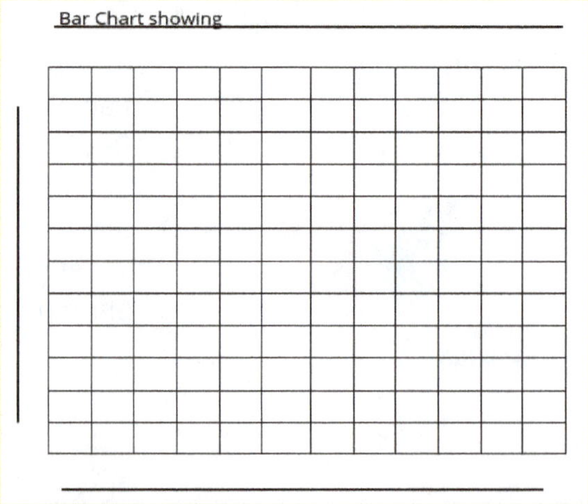

4. Pie chart of children who could correctly identify traditional Jamaican snacks from a group of 100 children.

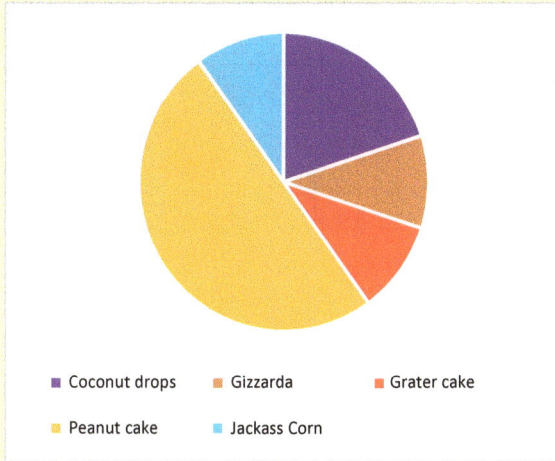

- Coconut drops ■ Gizzarda ■ Grater cake
- Peanut cake ■ Jackass Corn

Bar Chart showing _____

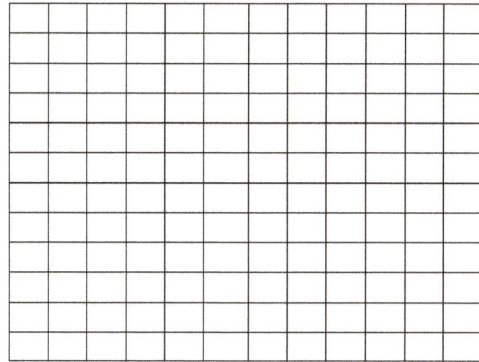

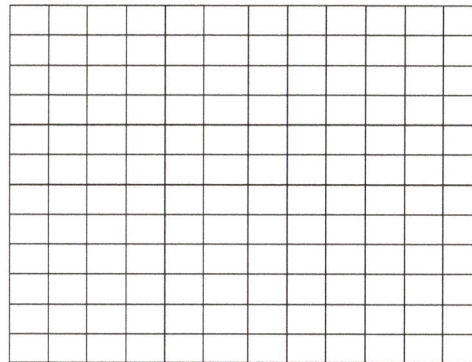

5. Pie chart showing how Janeil's mother spent her salary of $100 for the week.

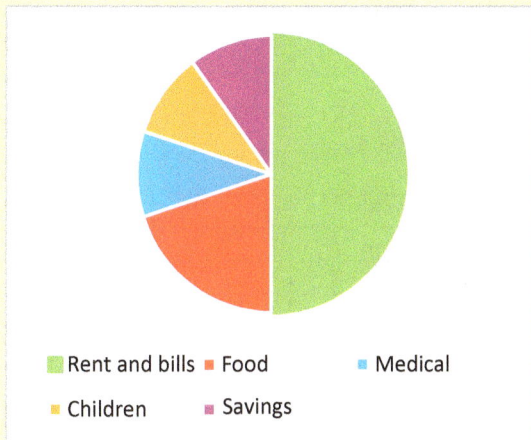

■ Rent and bills ■ Food ■ Medical
■ Children ■ Savings

Bar Chart showing _____

20.3 Convert a pictograph into bar graph or vice versa

Example: Conversion of Pictograph to Bar graph

Pictograph showing number of people vaccinated for one week at a clinic.

Days of the week	Monday	Tuesday	Wednesday	Thursday	Friday
Number of people vaccinated	👤👤👤👤👤	👤👤	👤👤👤	👤👤	👤👤👤👤👤👤👤👤

Each 👤 represents 5 people.

Solution: First create a data table from the pictograph

Days of the week	Monday	Tuesday	Wednesday	Thursday	Friday
Number of people vaccinated	25	10	15	10	40

Next draw the bar graph on the grid:

Bar chart showing number of people vaccinated for one week at a clinic

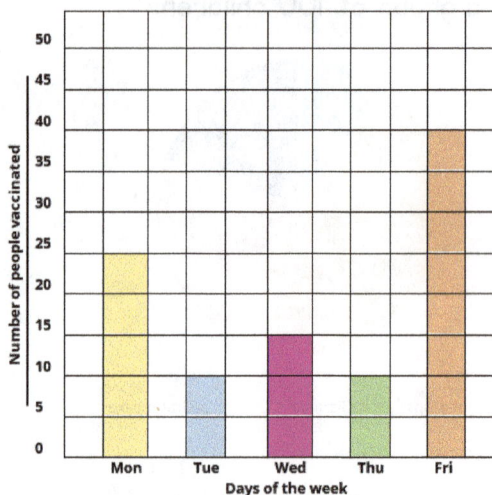

Let us practice 20.3.1

Create a Bar Chart from the following Pictographs.

1. Pictograph of grasshoppers seen at different places.

Bar Chart showing _____

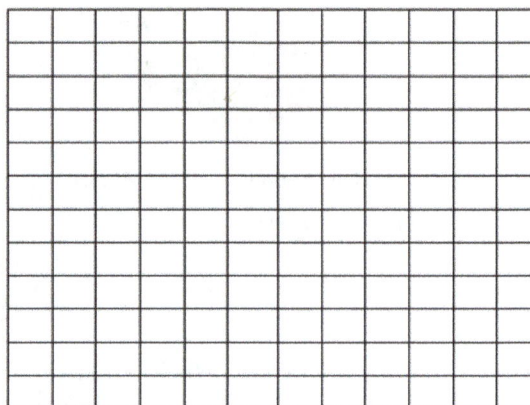

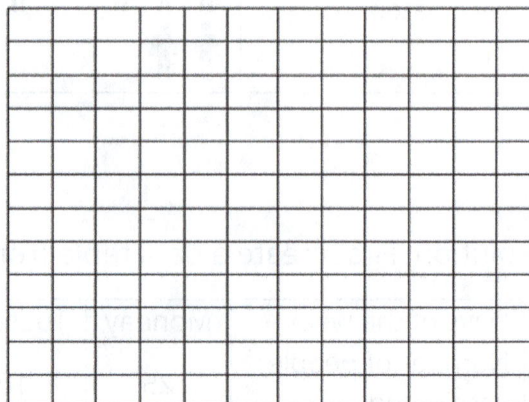

Places	Number of grasshoppers seen
School	🦗 🦗 🦗 🦗
Church	🦗
Home	🦗 🦗 🦗
Neighbour's yard	🦗 🦗 🦗 🦗 🦗 🦗
Park	🦗 🦗 🦗 🦗

Key: Each 🦗 = 4 grasshoppers

2. Pictograph showing the number of apples the five boys kept for themselves from the basket of apples.

Bar Chart showing _____

Boys	Number of apples taken
Dwayne	🍎 🍎 🍎 🍎
Devin	🍎 🍎 🍎
Desmond	🍎 🍎
Daniel	🍎 🍎 🍎 🍎
Dave	🍎 🍎 🍎 🍎 🍎

Key: Each 🍎 = 2 apples

3. Pictograph of shoes sold.

Shoe Stores	Number of shoes
Happy Feet	
The Shoe Shop	
Discount Centre	
Everything Shoe	
Walk in Style	

Key: Each = 10 shoes

Bar Chart showing _____

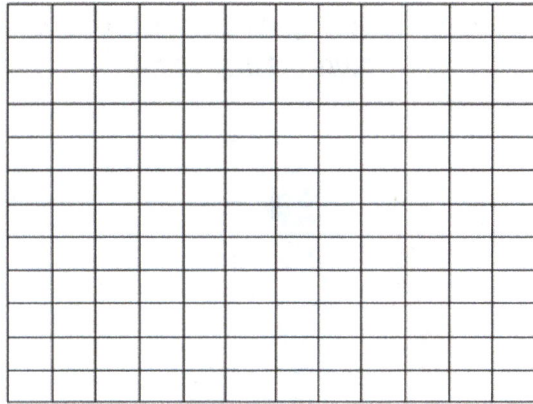

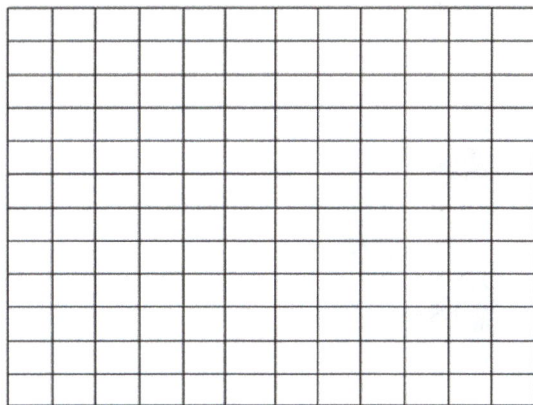

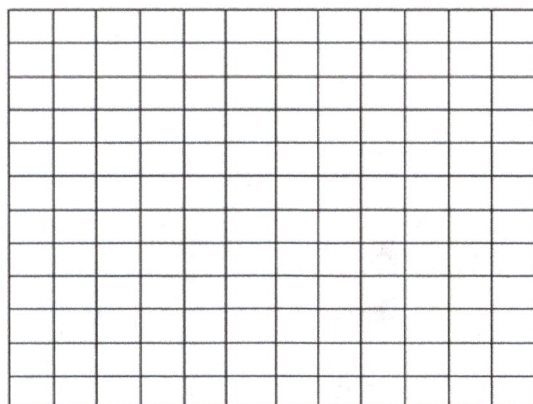

4. Pictograph of movies grade 4 children watched.

Movie watched	Number of children
Happy Feet	
Little Mermaid	
Dragon's Tales	
Despicables	
Little Critters	

Key: Each = 2 children

Bar Chart showing _____

5. Pictograph of cars parked in five parking lots.

Parking lots	Number of cars parked in each lot
A	
B	
C	
D	
E	

Key: Each = 4 cars

Bar Chart showing _____

Example of conversion of bar graph to pictograph.

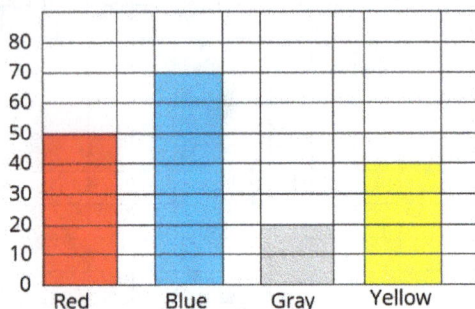

Bar chart showing the favourite colours of students in grade 4.

Pictograph of favourite colours of students in grade 4.

Colours	Number of students
Red	
Blue	
Gray	
Yellow	
Key: Each colour block = 10 student votes	

Let us practice 20.3.2

Use the bar charts to complete the pictographs.

1. In a survey, 40 students in a grade 4 class were asked what pets they had at home. The bar chart shows the number and types of pets the children have at home.

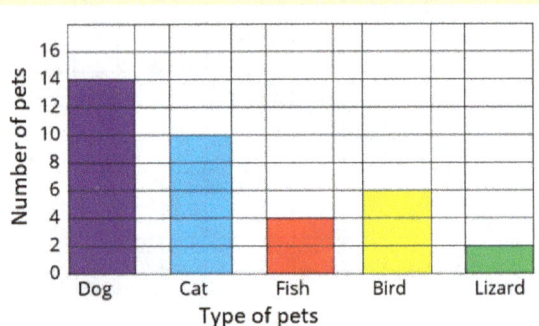

Pictograph of types of pets a grade 4 class have at home.

Animals	Pets
Dog	
Cat	
Fish	
Bird	
Lizard	

Key: Each animal = _____animals

2. Bar chart showing the number of computers sold at each of five shops at a technology fair.

Title: _____

Shops	Computers sold
1	
2	
3	
4	
5	

Key: Each computer represents ___ computers.

3. Bar chart showing the number of students in a class who were born in each month.

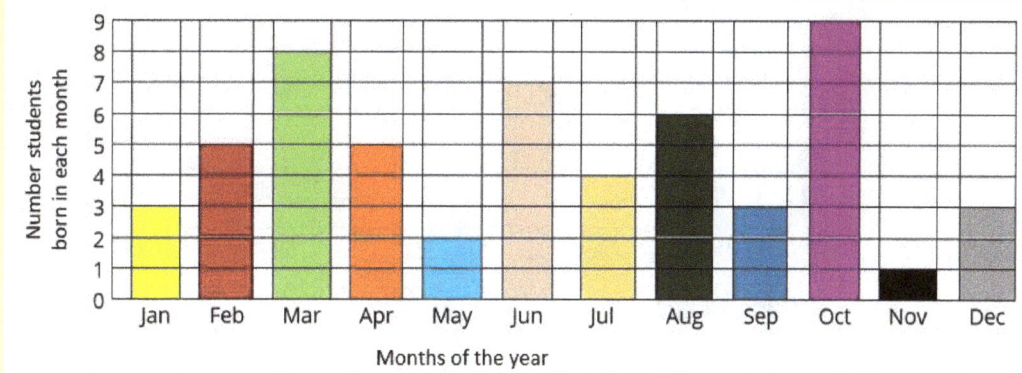

Title: _____

Month	Students	Month	Students
January	🧍🧍🧍		

Key: Each person = ____   _____

4. Bar chart of student participation in clubs in a grade 4 class.

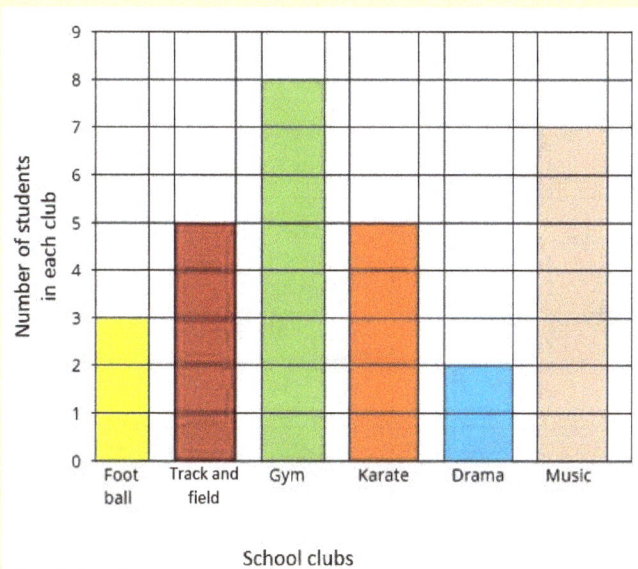

Title: _____

Clubs	Student participation

Key: Each coloured block = _____
_____

CHEETAH™
Connect to Higher Education, Electronic Tools, Aplication and Help

5. Bar chart of types of animal seen at a farm.

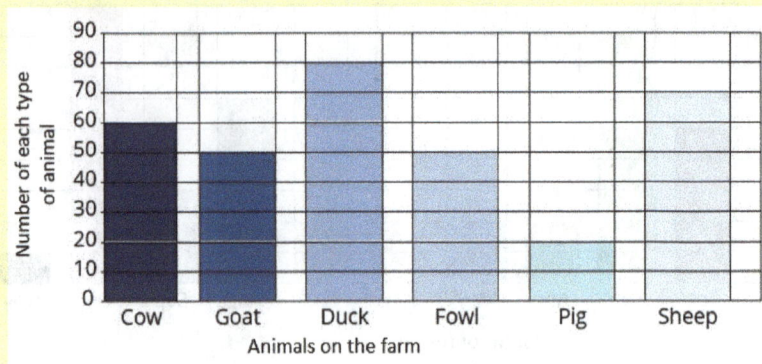

Title:_____

Animals	Number on a farm	Animals	Number on a farm

Key: Each animal = _____

Evaluation: Let us see how you did.

Learning outcome	No!	Working on it	Yes!
Did you get all the answers?	😞	😐	🙂
Did you get most questions right?	😞	😐	🙂
Did you retry the question(s) you got wrong?	😞	😐	🙂
Were you able to correct your wrong answers?	😞	😐	🙂
If not, did you seek help from others and/or review the chapter?	😞	😐	🙂

Colour the face that shows how you are doing.

TERM 3

Focus question

TERM 3, UNIT

How do I use my calculator to determine and prove results?

Chapter

Before we begin, let's see what you know.

Prior learning:

✓ Differentiate between the uses of the various operations in problem situations.
✓ Write pairs of multiplication and division facts from an array or given product and factors.
✓ Write story problem and solve.
✓ Use estimation in problem solving.

Key vocabulary

Check the words you understand:

☐ calculator
☐ dividend
☐ divisor
☐ leftovers
☐ mixed number
☐ patterns
☐ quotient
☐ relationships
☐ remainder

5. Write mathematical operation (add (+), subtract (-), divide (÷) or multiply (x)) needed for the situations listed. You may use more than one operation.

	Situation	Operation
a.	The truck moved 80 bags of sugar on the first day and 74 bags on the next day. How many bags of sugar were moved in the two days?	
b.	The distance from Town A to Town B is 54 kilometers. The taxi has gone 27 kilometers. What distance is left to reach Town B?	
c.	A man saved $76 every day. How much money does he save in five days?	
d.	The group needs to save $75 to buy pencils. If there are 5 students in the group, how much should each student give?	
e.	Mary bought 2 pens for $40 and a book for $70. She had only $100. How much more does she need to cover the bill?	
f.	John bought 2 pens for $30 **each** and two books for $70 **each**. He paid with $500 how much change does he receive?	

2. Multiplication facts or an array are the pairs of numbers that multiply to give the number of blocks in the array. Division facts are the factors of the number of blocks which make the array. Use these definitions to find multiplication and division facts for these arrays.

   a. Circle the multiplication facts for this array.

   1 x 24, 12 x 2, 3 x 5, 6 x 4, 3 x 4, 8 x 3

   b. Which of these numbers are factors (division facts) for the total number of small blocks in the array? Circle your answers.

   1   2   3   4   5   6   7   8

c. Can you tell how many small squares are in the array without counting each?

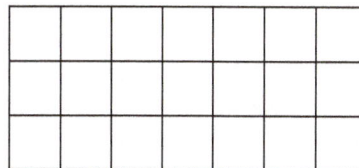

d. Write multiplication facts for the array.

Which of these numbers could not be a factor of the number of blocks? Circle your answer(s).

1   2   3   4   5   6   7   8

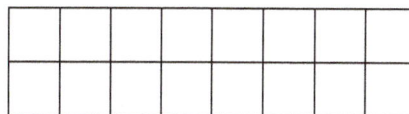

e. Write how many small blocks are in the array without counting each. _____

Which of these numbers could not be a factor of the number of blocks in the array? Circle your answer(s).

1   2   3   4   5   6   7   8

3. Use the array to write, how many people would eat chocolate, if each person gets 3 blocks. How many blocks of chocolate would be left over?

Look at these story questions and solve them.

1. A farm has 8 cows. Each cow gives 6 litres of milk daily. How much milk is collected on the farm each week? _____

2. A car maker makes 7 cars each day. At the end of a 28-day month how many cars are made? ____

3. A man drank 6 litres of water each day. How many days does it take him to drink 54 litres? ____

4. The jug had 24 litres of juice. After 8 people were served only 9 litres were left in the jug. How many litres did the people drink in all? _____

5. On Monday Leon answered 24 phone calls, Beverly answered 14 more than Leon answered. How many calls were answered in all? _____

6. At the clinic the doctor had 42 appointments for the day but only 27 people came to see the doctor. How many people missed their appointment? _____

7. There are 57 students in grade 4. If 19 students are late for school how many students are early for school? _____

8. There are 268 people at the concert. There are 89 males, 104 females. How many children are at the concert? _____

9. Mary took 27 minutes more than John to finish the job. John took 38 minutes. How many minutes did Mary take to finish the job? _____

10. The school auditorium had 185 seats. 234 parents came to the meeting. How many at the meeting had no seat? _____

If you know these, you should be able to learn what comes next.

21.1   Define and use the terms dividend, quotient, divisor, remainder in sentences requiring division.
21.2   Divide numbers of up to five digits by numbers up to two digits, with or without remainder.
21.3   Divide a 3, 4 or 5 digit number so that zero is a digit in the tens and/or hundreds place in the quotient.
21.4   Test for divisibility by 2, 3 or 4.
21.5   Express, as a mixed number, the answer to a division problem with a remainder.
21.6   Identify and correct wrong answers in problems involving division.
21.7   Discover, memorize and recall all division facts up to at least 1010100 .
21.8   Identify and use the keys on a pocket calculator.
21.9   Use the calculator to check answers.   21.10 Investigate number patterns using the calculator.

21.1 Define and use the terms dividend, quotient, divisor, remainder in sentences requiring division.

The long division shows the **relationship** between the **divisor**, the **dividend**, the **quotient** and the **remainder**.

See that 43 can divide 3 into 14 equal parts with 1 as a remainder. So, $43 \div 3 = 14\frac{1}{3}$

Also, note that the divisor x quotient + the remainder = dividend.
When there is no remainder, the divisor x the quotient is the dividend.
And the quotient is equal to the dividend divided by the divisor.

Let us practice 21.1

1. What is the quotient of 78 divided by 6?	2. What is the remainder of 358 divided by 5?
3. What is the dividend if the quotient is 8, the divisor is 9 and there is no remainder?	4. What is the divisor if the dividend is 168 and the quotient is 12?
5. What is the dividend if the quotient is 24, the divisor is 8 and the remainder is 3.	6. There are 300 nails in a box and each photo frame needs 8. How many frames can be built and how many nails are not used?
7. The biscuits cost $371. i. If 8 people paid for biscuits, how much did each pay? ii. How much change was left?	8. Nine students are to collect $860. If each person gives $90, how much more money is needed?
9. A book has 260 pages. Sam reads 8 pages every night. How many pages will he read on the last night?	10. A teacher shares 189 balloons equally among 5 students to decorate the classroom. How many balloons will she have left?

21.2 Divide numbers of up to five digits by numbers up to two digits, with or without remainder.

Example 1: 3 digits divided by 1 digit

$$9 \overline{\smash{)}808}$$

89

−72
88
−81
7

9 into 80 = 8
8 x 9 = 72
80 – 72 = 8
Bring down the 8 = 88
9 into 88 = 9
9 x 9 = 81
88 – 81 = 7
Remainder is 7

Example 2: 4 digits divided by 2 digits

248

$$12 \overline{\smash{)}2987}$$

−24
58
−48
107
−96
11

12 into 29 = 2
2 x 12 = 24
29 – 24 = 5
Bring down the 8 = 58
12 into 58 = 4
4 x 12 = 48
58 – 48 = 10
Bring down the 7
12 into 107 = 8
8 x 12 = 96
107 - 96 = 11
Remainder is 11

Example 3: 5 digits divided by 2 digits

876

$$13 \overline{\smash{)}11198}$$

−102
99
−91
88
−78
10

14 into 111 = 8
8 x 13 = 102
111 – 102 = 9
Bring down 9 = 99
13 into 99 = 7
7 x 13 = 91
99 – 91 = 8
Bring down 8 = 88
13 into 88 = 6
6 x 13 = 78
88 - 78 = 10
Remainder is 10

Let us practice 21.2

Solve the following questions.

1. 876 ÷ 4

2. 705 ÷ 5

3. 626 ÷ 7

4. 3563 ÷ 8

5. 6768 ÷ 9

6. 815 ÷ 10

7. 1452 ÷ 11

8. 7914 ÷ 12

9. 6372 ÷ 13

10. 560 ÷ 14

11. 4673 ÷ 18

12. 3525 ÷ 25

CHEETAH
Connect to Higher Education, Electronic Tools, Aplication and Help

21.3 Divide a 3, 4 or 5 digit number so that zero is a digit in the tens and/or hundreds place in the quotient.

Examples:

Zero in the tens place of the quotient	Zero in the tens place of the quotient	Zero in the hundredth place of the quotient
636 ÷ 6	3002 ÷ 5	8597 ÷ 8
$$\begin{array}{r} 106 \\ 6\overline{)636} \\ -6\downarrow \\ \hline 3\downarrow \\ -06 \\ \hline 36 \\ -36 \\ \hline 0 \end{array}$$	$$\begin{array}{r} 600 \\ 5\overline{)3002} \\ -30 \\ \hline 00 \\ -00 \\ \hline 02 \\ -00 \\ \hline 2 \end{array}$$	$$\begin{array}{r} 1074 \\ 8\overline{)8597} \\ -8 \\ \hline 5 \\ -0\downarrow \\ \hline 59 \\ -56\downarrow \\ \hline 37 \\ -32 \\ \hline 5 \end{array}$$

Let us practice 21.3

1. Find the quotients.
   a. 2149 ÷ 7
   b. 1236 ÷ 6
   c. 8360 ÷ 8
   d. 18585 ÷ 9

   e. 2024 ÷ 4
   f. 4808 ÷ 8
   g. 5663 ÷ 7
   h. 72608 ÷ 8

2. Find the quotients.
   a. 750 ÷ 5
   b. 4920 ÷ 6
   c. 650 ÷ 13
   d. 3150 ÷ 15
   e. 7700 ÷ 11

   f. 8463 ÷ 21
   g. 1836 ÷ 18
   h. 14472 ÷ 24
   i. 16.215 ÷ 23
   j. 10556 ÷ 26

21.4 Test for divisibility by 2, 3 or 4.

Divisibility rules	
Divisible by 2	All numbers ending with 0, 2, 4, 6 or 8 are divisible by 2. Numbers ending with an even number (2, 4, 6, or 8) or 0 are divisible by 2. Example 1: 806 ÷ 2 = 403 hence 806 is divisible by 2. Example 2: 753 ÷ 2 = 188.75, hence 753 is not divisible by 2.
Divisible by 3	A number is divisible by 3, If the sum of the digits is divisible by 3. Example 1: For 24, the sum is 4 + 2 = 6 and $\frac{6}{3}$ = 2 So, 24 is divisible by 3. Example 2: For 8642, the sum is 8 + 6 + 4 + 2 = 20) and $\frac{20}{3}$ = 6.66... So, 8642 is not divisible by 3.

Divisible by 4	A number is divisible by 4, if the last 2 digits of the number is divisible by 4 or if the last 2 digits of the number is 00.
	Example 1: For 724; 24÷ 4 = 6, hence 724 is divisible by 4.
	Example 2: For 354; 54 ÷ 4 = 13.5, hence 354 is not divisible by 4.
	Example 3: For 500; 500 ÷ 4 = 125.

Let us practice 21.4

1. Look at the numbers in the table, then record which are divisible by 2, 3 or 4.

8, 13, 24, 66, 128, 150, 184, 472, 592, 502, 594, 406, 472, 506, 358, 48, 78, 394		
a.	2	
b.	3	
c.	4	

2. Look at the numbers and tick the numbers that are divisibile by 2, 3, 4 or neither of each.

Item	Number	2	3	4	None
a.	24				
b.	96				
c.	48				
d.	72				
e.	763				
f.	1104				
g.	153				
h.	203				
i.	748				
j.	852				

21.5 Express, as a mixed number, the answer to a division problem with a remainder.

Expressed as mixed fractions	Long divisions
$26 \div 8 = \frac{26}{8} = 3\frac{1}{4}$	$$\begin{array}{r} 3 \\ 8\overline{)26} \\ -24 \\ \hline 2 \end{array}$$
$47 \div 5 = \frac{47}{5} = 9\frac{2}{5}$	$$\begin{array}{r} 9 \\ 5\overline{)47} \\ -45 \\ \hline 2 \end{array}$$
$1125 \div 12 = \frac{1125}{12} = 93\frac{3}{4}$	$$\begin{array}{r} 93 \\ 12\overline{)1125} \\ -108 \\ \hline 45 \\ -36 \\ \hline 9 \end{array}$$

CHEETAH™
Connect to Higher Education, Electronic Tools, Aplication and Help

Let us practice 21.5

Solve each problem using long division and express the answer as a mixed number.

a. $9846 \div 7$

b. $4672 \div 5$

c. $3675 \div 8$

d. $9275 \div 9$

e. $5673 \div 4$

f. $1127 \div 3$

g. $8249 \div 11$

h. $2745 \div 10$

i. $8656 \div 12$

21.6 Identify and correct wrong answers in problems involving division.

Some common errors in division include:
- incorrect position of the quotient
- incorrect multiplication
- incorrect position of quotient x divisor result
- bringing down the incorrect number
- omitting the remainder
- incorrect subtraction
- remainder larger than divisor at end of long division
- incorrect division to find the quotient.

Let us practice 21.6

Identify the errors in each problem and rewrite the correct solution.

1. $1234 \div 3$

```
      219
  6 ) 1234
     -12
      11
     - 6
      54
     -54
       0
```

2. $347 \div 4$

```
       94
  4 ) 347
     -32
      27
    - 28
       1
```

3. $8762 \div 7$

```
     1341
  7 ) 8762
    -7
     17
    -14
     36
    -35
     12
     -7
      6
```

4. $3891 \div 4$

```
     0983
  4 ) 3891
    -36
     29
    -28
     11
    -12
     ..
```

5. 379 ÷ 3

```
        139
    3 ⟌ 379
       -3
        7
       -9
        29
       -27
        2
```

6. 4835 ÷ 7

```
         9176
     7 ⟌ 4835
        -36
         12
         -7
         53
        -49
         45
        -42
         3
```

Evaluation: Let us see how you did.

Learning outcome	No!	Working on it	Yes!
Did you get all the answers?	☹	😐	🙂
Did you get most questions right?	☹	😐	🙂
Did you retry the question(s) you got wrong?	☹	😐	🙂
Were you able to correct your wrong answers?	☹	😐	🙂
If not, did you seek help from others and/or review the chapter?	☹	😐	🙂

Colour the face that shows how you are doing.

Before we begin, let's see what you know.

Prior learning:
✓ Identify and name plane shapes.

Plane shapes also called flat shapes, are closed two-dimensional figures made by joining curved or straight lines. Some examples of plane shapes are squares, rectangles, circles and triangles. Different plane shapes have different properties, such as the number of sides or corners (or vertices).

Draw lines to match the following plane shapes with their names.

Plane shape	Name
	hexagon
	octagon
	circle
	oval
	rectangle
	triangle
	square
	decagon
	pentagon

Key vocabulary

Check the words you understand:

☐ cross section
☐ cube
☐ cuboid
☐ edge
☐ face
☐ net
☐ prism
☐ vertex

If you know these, you should be able to learn what comes next.

22.1 Develop/create skeletons/frames of solid shapes using a variety of tools.
22.2 Draw and describe nets of prisms: cubes and cuboids.
22.3 Construct solids from given nets (prisms: cubes and cuboids)
22.4 Identify and explore the properties of prisms: cubes and cuboids.
22.5 Identify prisms from their nets: cubes and cuboids

22.1 Develop/create skeletons/frames of solid shapes using a variety of tools.

Solid shapes and objects are different from two-dimensional (2D) shapes and objects because of the presence of the 3 dimensions—length, breadth, and height. Because of these dimensions, objects have faces, edges and vertices.

Examples of solid shapes as real-life three-dimensional objects are seen in every object, be it a cup, ball, an ice cream cone, a box, or television.

Properties of a cube

Consider a cube of ice in the tray of your fridge. All six faces of the cube are the same, which also makes it a square prism. Even a Rubik's Cube or a playing die are examples of cubes. Here are some truths about a cube.

1. It has all the faces in the shape of a square.
2. All the faces or sides of a cube have equal dimensions.
3. The angles of planes of the cube are the right angle.
4. Each of the faces of a cube meet the other four faces.
5. Each of the vertices of a cube meet the three faces and three edges.
6. The edges that are opposite to each other are parallel.

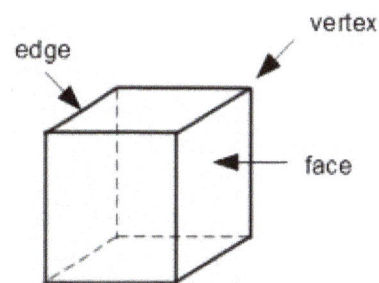

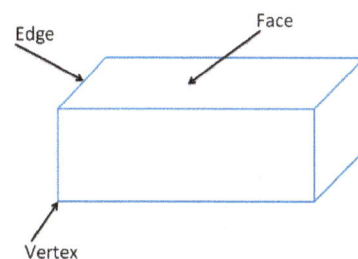

Properties of a Cuboid

A cuboid is like a matchbox or a book. Here are some characteristics of a cuboid.

1. It has all the faces in the shape of a rectangle.
2. All the faces or sides of a cuboid have different dimensions.
3. The angles of planes of the cuboid are the right angle.
4. Each of the faces of a cuboid meet the other four faces.
5. Each of the vertices of a cuboid meet the three faces and three edges.
6. The edges that are opposite to each other are parallel.

Let us practice 22.1

1. Use the properties of cubes and cuboids given to complete the table.

Shape	No. of edges	No. of vertices	No. of faces	Shape of faces
cube				
cuboid				

2. Identify and change the incorrect information in this table.

Shape	No. of edges	No. of vertices	No. of faces	Shape of faces
cube	12	8	8	rectangle
cuboid	9	6	7	square

3. Select the correct label.

	I	II	III
A	face	edge	vertex
B	edge	vertex	face
C	vertex	face	edge
D	vertex	edge	face

22.2 Draw and describe nets of prisms: cubes and cuboids.

A net is the pattern made when the surface of a three-dimensional figure is laid out flat showing each face of the figure. A net is folded to make a three-dimensional figure.

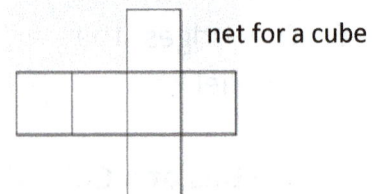

net for a cube

Consider that you were asked to build two boxes by folding paper: one a cube and the other a cuboid. To build the boxes you would probably come up with drawings or cut out and fold these patterns. These cut-outs are the nets of a cube and cuboid.

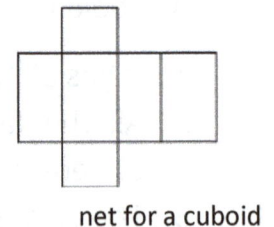

net for a cuboid

Let us practice 22.2.1

Many other shapes would fold to give the same results.

Draw, cut out and fold these nets and decide which will give a cube.

1. A.   B.

2. A.   B.

3. A.   B.

4. A.   B.

Let us practice 22.2.2

1. Many other shapes would fold to give the same results. Draw, cut out and fold these nets to decide which will give a cuboid.

i.	ii.	iii.
iv.	v.	vi.
vii.	viii.	ix.
x.	xi.	xii.

Evaluation: Let us see how you did.

Learning outcome	No!	Working on it	Yes!
Did you get all the answers?	☹	😐	🙂
Did you get most questions right?	☹	😐	🙂
Did you retry the question(s) you got wrong?	☹	😐	🙂
Were you able to correct your wrong answers?	☹	😐	🙂
If not, did you seek help from others and/or review the chapter?	☹	😐	🙂

Colour the face that shows how you are doing.

How do I use variables when solving real world problems?

Before we begin, let's see what you know.

Prior learning:
- ✓ Write number sentences in words.
- ✓ Know how to use the four basic operations.
- ✓ Use symbols to represent unknown numbers.

Key vocabulary

Check the words you understand:

- ☐ algebraic
- ☐ compute
- ☐ constant
- ☐ expression
- ☐ formulae
- ☐ operations
- ☐ patterns
- ☐ solution
- ☐ substitution
- ☐ variable

1. If A is less than 6, but greater 4, which of these numbers could be A?

   a. 7　　　　　b. 6　　　　　c. 5　　　　　d. 2

2. Draw lines to match the number sentences with their word sentence.

Number sentences	Word sentences
7 + 19 = 26	If the sum of 2 numbers is 8 and one of the numbers is 6, what is the other number?
2 + 8 =10	There are 40 students in the class and 28 came early. How many came late?
25 – 9 = 16	Increase 7 by 19 to give 26.
18 ÷ 6 = ?	How many times is 18 more than 6?
24m ÷ 6 = ?	There are 8 benches in the class and each bench holds four students. How many students are in the class?
a + b = 8 and a = 6, Find b	Decrease 25 by 9.
12 ÷ a = 4 find a	A piece of thread 24 metres long was cut into 6 equal parts. How long was one part?
8 x 4 = ?	Two plus 8 is 10.
40 – 28 = ?	When 12 apples were shared equally each child got four. How many children got apples?

If you know these, you should be able to learn what comes next.

23.1　Identify the correct operation to be used in solving a problem.
23.2　Solve word problems using algebraic expressions.
23.3　Demonstrate the principle of substitution in simple formulae.

23.1 Identify the correct operation to be used in solving a problem.

The four mathematical **operations** are addition, subtraction, multiplication and division. Look at the problem. When solving problems, we apply the rule of BOMDAS orders (bracket, multiplication, division, addition and subtraction), which tells what mathematical operation to use first.

BOMDAS means **B**racket **O**f **M**ultiplication **D**ivision **A**ddition **S**ubtraction

$2 \times 3 + 1 = ?$ Is it $6 + 1 = 7$ or $2 \times 4 = 8$?

When BOMDAS is used, brackets are removed first, then multiplications are done, followed by division, then addition then subtractions.

Brackets () first means: $3(4) = 3 \times 4 = 12$

$$3(5-1) = 3 \times 5 - 3 \times 1 = 15 - 3 = 12$$

$$3(5-1) = 3 \times 4 = 12$$

This is known as removing the brackets.

Example 1: Simplify $3 \times 4 \div 2 + 5$

Multiply first ($3 \times 4 = 12$)	$= 12 \div 2 + 5$
Divide product by 2 ($12/2 = 6$)	$= 6 + 5$
Add 5 ($6 + 5 = 11$)	$= 11$

Example 2: Simplify $2(3 + 2) - 6$

Note, what is outside the bracket multiplies everything inside the bracket.

Work inside the brackets first to give	$= 2(5) - 6$
Remove brackets to give	$= 2 \times 5 - 6$
	$= 10 - 6$
Simplify to give	$= 4$

$8 - 5 + 4$ is the same as $8 + 4 - 5$ is the same as $3 + 4$

Example 3: $3(4 - 2) + 8 \div 4$

Work what is inside the brackets first, $4 - 2 = 2$ to get	$= 3(2) + 8 \div 4$
Remove brackets to get	$= 3 \times 2 + 8 \div 4$
Work multiplication to get	$= 6 + 2$
Simplify to get	$= 8$

Let us practice 23.1

Solve the following problems.

1. $5 + 2 - 3$
2. $7 - 5 + 6$
3. $4 + 3 \times 2$
4. $2 \times 3 - 2$

5. $9 - 2 + 4$
6. $5(3) - 4$
7. $6 \div 2 + 7$
8. $9 - 8 \div 4$

9. $2 + 3 \times 4 - 1$
10. $9 - 4 \div 2 + 1$
11. $3(5 + 2)$
12. $4(6 - 1)$

13. $(7 - 4) + 2$
14. $(4 + 2) - 2$
15. $4 + (2 - 2)$

23.2 Solve word problems using algebraic expressions.

Word problems can be solved by changing them to an **algebraic expression**.

Here is an example: *Tom is 2 years older than John and the sum of their ages is 16 years. What are their ages?*

To solve, let the age of John be shown as the unknown number x.

If John is x years old, then Tom is (x+2) years old. So, the algebraic expression is (x+2) + x.

If the sum of their ages is 16, this means that (x+2) + x = 16 which can be solved.

$(x+2) + x = 16$

$x + 2 + x = 16$	remove the bracket
$x + x + 2 = 16$	bring all unknowns (x) together
$2x + 2 = 16$	add all the xs together
$2x-2 = 16 -2$	getting ready to leave x alone
$2x = 14$	take away 2 from both sides
$\frac{2x}{2} = \frac{14}{2}$	to get x by itself, we divide both sides by 2

$x = 7$

Therefore, John is 7 years old, while Tom is 7 + 2 = 9 years old.

Let us practice 23.2

1. Isaiah is 3 times as old as Craig and Craig is 4 years old. How old is Isaiah? _____
2. Timothy is 3 years older than James. James is 5 years old. How old is Timothy? _____
3. Elaine has sheets of coloured paper. She gave 55 sheets to Jane. She now has 135 sheets left. How many sheets of paper did Elaine have before? _____
4. Bianca has sweets. She gave 25 to her friend Teon, and twice as many to Alexia. If she has 86 left, how many did she have at first? _____
5. When a number (x) is increased by 8 the answer is 24. What is the number? _____
6. When p is added to 8 the answer is 20. What is p? _____
7. Five (5) times a number gives 80. What is the number? _____
8. Half of a number is 12. What is the number? _____
9. Twenty-four (24) divided by a number gives 6. What was the number? _____
10. A number when multiplied by itself gives 36. What is the number? _____
11. Double a number, then find two times the answer and you have 30. What is the number? _____

23.3 Demonstrate the principle of substitution in simple formulae.

Remember that **substitution** is replacing one thing with another. If you have the simple **formula** a = 3b, you can replace the b with the number 4 like this:

a = 3b

a = 3 x b

a = 3 x 4

a =12.

> Substitution allows us to replace the letter or symbol we do not know to solve simple formulae.

Given that ab = a x b. If a = 2, b = 3, c = 4, d = 5, e = 6, f = 7.

Example 1:	**Example 2:**	**Example 3:**
2a + 2b + 2c	3d – 2a + 2c	b x e x d
= 2 x 2 + 2 x 3 + 2 x 4	= 3 x 5 – 2 x 2 + 2 x 4	= 3 x 6 x 5
= 4 + 6 + 8	= 15 – 4 + 8	= 3 x 30
= 18	= 11 + 8	= 90
	= 19	
Example 4:	**Example 5:**	**Example 6:**
a + b +c +d +e -f	ef - dc	$\left(\frac{e}{a}\right) - 1$
= 2 + 3 +4 +5 +6 - 7	= 6 X 7 – 5 X 4	$= \left(\frac{6}{2}\right) - 1$
= 20 - 7	= 42 - 20	= (3) – 1
= 14	= 22	= 3 – 1
		= 2

Let us practice 23.3.1

Given a = 2, b = 3, c = 4, d = 5, solve the following.

1. 2b

2. Cab

3. 3a + 2c + 2b – 2d

4. 4a + 4b – 4c

5. 2d + 2c + 2b + 2a

6. 7a + 6b + 5c + 4d

Substituting missing numbers into algebraic formulae.

Example 1: If R + R = 40, R + P = 23, P + T = 10, find R + P + T

If R + R = 40,	R + P = 23	And if P + T = 10	Therefore
2R = 40	20 + P = 23	3 + T = 10	R + P + T
$R = \frac{40}{2} = 20$	P = 23 - 20	T = 10 - 3	= 20 + 3 + 7
	P = 3	T = 7	= 30

CHEETAH

Connect to Higher Education, Electronic Tools, Aplication and Help

Example 2: If $\square + \square = 14$, $\square + \stackrel{\star}{} = 16$, $\stackrel{\star}{} + \bigcirc = 24$,

Find $\square + \stackrel{\star}{} + \bigcirc = ?$

If $\square + \square = 14$	$\square + \stackrel{\star}{} = 16$	$\stackrel{\star}{} + \bigcirc = 24,$	$\square + \stackrel{\star}{} + \bigcirc = ?$
$2\square = 14$	$7 + \stackrel{\star}{} = 16$	$9 + \bigcirc = 24,$	$= 7 + 9 + 15$
$\square = \frac{14}{2}$	$\stackrel{\star}{} = 16 - 7$	$\bigcirc = 24 - 9$	$= 31$
$\square = 7$	$\stackrel{\star}{} = 9$	$\bigcirc = 15$	

Let us practice 23.3.2

1. If $x + x = 6$; $x + y = 9$, $y + z = 14$; Find $x + y + z$

2. If $A + A = 4$; $A + B = 9$, $B + C = 14$; Find $A + B + C$

3. If 🍾 + 🍾 = \$ 50, 🍾 + 🍔 = \$ 75, 🍔 + 🍟 = \$ 80

Find 🍾 + 🍔 + 🍟 = ?

4. When the hangar is level, the total mass of the weights on the hanger is 36 g. If the masses of two of the weights are 8 g and 3 g as shown, what are the masses of the other 2 shapes, ▲ and ▌ ?

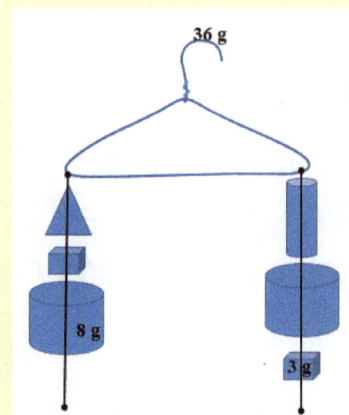

5. If + = \$10; + = \$15; + = \$17; Find + −

6. If + = \$8; + = \$9; + = \$11; Find x +

7. If + = \$20; + = \$19; + = \$14; Find + −

8. If + = \$40; + = \$70; + = \$90;

Find + +

Evaluation: Let us see how you did.

Learning outcome	No!	Working on it	Yes!
Did you get all the answers?	😞	😐	🙂
Did you get most questions right?	😞	😐	🙂
Did you retry the question(s) you got wrong?	😞	😐	🙂
Were you able to correct your wrong answers?	😞	😐	🙂
If not, did you seek help from others and/or review the chapter?	😞	😐	🙂

Colour the face that shows how you are doing.

Prior learning:
✓ Use probability terms accurately.

Before we begin, let's see what you know.

1. Which 2 of these words best describe the word probability?

    a. certainty   b. option   c. chance

    d. guess   e. likelihood   f. possible

If you know this, you should be able to learn the following.

Key vocabulary

Check the words you understand:

☐ chance   ☐ drag
☐ experience   ☐ flight
☐ possible outcome
☐ prediction
☐ probability ☐ pull
☐ push ☐ resistance
☐ sample space ☐ thrust

🎯 24.1 Make predictions regarding the outcomes of experiments and record the results explaining any differences.
24.2 Predict and record the likely outcome of an experiment.

24.1 Make predictions regarding the outcomes of experiments and record the results explaining any differences. 24.2 Predict and record the likely outcome of an experiment.

A prediction about something (an outcome) happening is the **chance** that something may happen, also called **probability**. How do you find the probability (chance) of something happening?

$$\text{Probability of something happening} = \frac{\text{number of times a selected event may happened}}{\text{total number of possible happenings}}$$

Probability is the chance something happens or how likely it is to happen. We use probability to guess what may happen in different situations.

Example 1: Look at a coin toss. How do we decide which side gets the ball first in a football, netball or cricket match? We toss a coin and ask the teams to predict if the coin will fall on heads or tails. Since the coin has only two sides, then the chance it will fall on either side is 1 out of 2 possible happenings.

Hence the probability of getting head in a coin toss is $\frac{1}{2}$ or 0.5.

The greatest chance of something happening is 1 and the least chance of something happening is 0. So, the answer for a probability is 0 ,1 or a proper fraction.

Example 2: *What is the chance of selecting a red ball from a bag containing only red balls?* The answer is 1, because if there are 5 balls and all are red, we get $\frac{5}{5} = 1$

Example 3: *What is the chance of selecting a blue ball from a bag containing only red balls?* The answer to this is 0, because if there are 5 red balls and 0 blue balls, we get $\frac{0}{5} = 0$

Example 4: *Look in this box containing numbers and letters written on the same-sized cards. Suppose you put your hand into the box, close your eyes and pick one card from the box. What is the chance that you could take up a number card from the box?*

I	A	B	3
C	D	4	E
7	8	9	F

Probability of taking up a number = $\frac{number\ of\ number\ cards}{total\ number\ of\ cards}$

Probability of taking up a number card = $\frac{6}{12} = \frac{1}{2}$

Example 5: *If a die is thrown onto a ludo table, you are able to read the top side when it lands. What is the probability or chance of a 6 showing on top?*

Since there are 6 sides and one 6, the probability of a 6 showing on top is:

$$\frac{1\ side}{6\ sides} = \frac{1}{6}$$

What is the probability or chance that a number larger than 3 shows on top?

Since there are 6 sides and 3 numbers larger than 3 (4, 5, and 6), the probability of a number greater than 3 showing on top is: $\frac{3\ sides}{6\ sides} = \frac{3}{6} = \frac{1}{2}$

Let us practice 24.1 & 24.2

1. Answer the following questions using the jar of coloured balls shown.
   i. What is the probability that a blue ball is selected from the jar? _____
   ii. What is the probability that a red ball is selected from the jar? _____
   iii. What colour ball has a 1/10 chance of being selected from the jar? _____

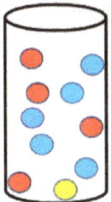

2. Answer the following questions using the jar of coloured balls shown.
   i. What is the probability that a green ball is selected from the jar? _____
   ii. What is the probability that a yellow ball is selected from the jar? _____
   iii. What is the probability that a round ball is selected from the jar? _____
   iv. What is the probability that a stripe ball is selected from the jar? _____
   v. What colour ball has a 1/13 chance of being selected from the jar? _____

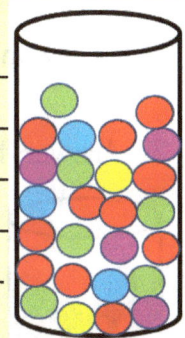

3. Answer the following questions using the die shown.
   i. How many different outcomes can you have when a die is rolled? _____
   ii. What is the probability that the die lands on three? _____
   iii. What is the probability that a four is selected when the die is rolled? _____

iv. What is the probability that the die lands on a number greater than four? _____

v. What is the probability that the die lands on a number less than two? _____

4. A canteen bought patties to be sold to the children at a school. Answer the questions using the graph.

i. What is the total number of patties bought? _____

ii. What is the probability that a patty bought is a chicken patty? ____

iii. What is the probability that a patty bought is a soy patty? _____

iv. What is the probability that a student gets a cheese patty? _____

Number of patties bought

Number of patties (y-axis: 0, 10, 20, 30, 40, 50, 60)

Beef 50, Chicken 20, Soy 10, Vegetable 10, Shrimp 10

Types of patties

v. If patties are randomly given to students, what is the most likely patty to get? Explain using the probability to get that patty. _____

vi. Getting which three types of patties would have an equal probability? _____

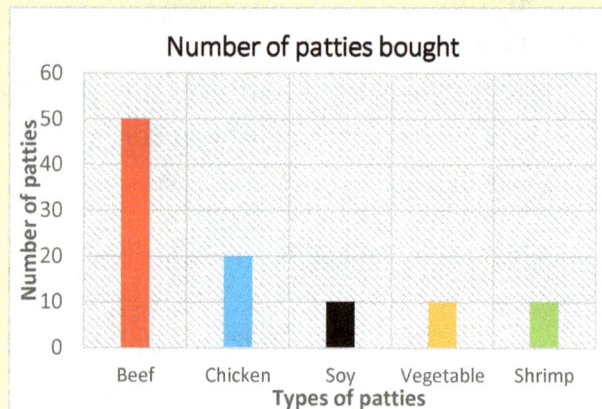

5. Using the die shown, how likely is it to roll the following?

i. more than 3 _____   ii. less than three _____

iii. an odd number _____   iv. a whole number _____

6. Answer the following questions using the chart shown.

i. What is the probability that Janeil's mother will spend her salary on rent and bills? _____

ii. How likely is it that Janeil's mother will spend her salary on medical expenses? _____

iii. What is the probability that Janeil's mother will not spend her salary on food? _____

iv. How likely is it that Janeil's mother will spend $10 on one of her 5 expenses? _____

Pie chart showing how Janeil's mother spent her salary of $100 for the week

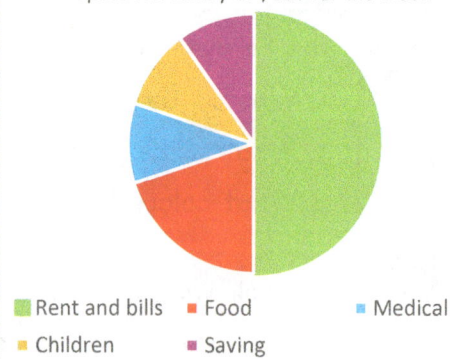

■ Rent and bills ■ Food ■ Medical
■ Children ■ Saving

7. Answer the following questions using the jar of marbles shown.

i. How likely are you to pick a blue marble from the jar?

a. not likely   b. very likely   c. certain   d. impossible

ii. What is the least likely marble to pick from the jar?

a. green   b. red   c. yellow   d. purple

iii. What colour ball has a 6/26 chance of being selected from the jar?

CHEETAH
Connect to Higher Education, Electronic Tools, Aplication and Help

8. Answer the following questions using a tally sheet of survey results on cars entering a school.

Colour cars	White	Red	Blue	Yellow	Black
Tally	⊮⊮⊮ ⊮⊮⊮ ⊮⊮⊮	⊮⊮⊮ III	II	⊮⊮⊮	⊮⊮⊮ ⊮⊮⊮

i. What is the probability that a red car will enter the school? _____

ii. How likely is it that a blue car will enter the school?
   a. least likely   b. impossible   c. most likely   d. certain

iii. From the answers given, what is the most likely coloured car to enter the school?
   a. blue   b. purple   c. yellow   d. red

iv. What colour car has a $\frac{1}{8}$ chance of entering the school? _____

9. There are 52 cards in a deck. Thirteen of the cards are spades, 13 are hearts, 13 are clubs and 13 are diamonds. What is the probability that a card chosen at random would be a spade? _____

10. In a board game, children were asked to use a spinner to choose what they would have for lunch.

   i. What is the probability that a child would get a burger for lunch? _____
   ii. What is the probability that a child would get watermelon for lunch? _____
   iii. What two foods would they have the same chance of getting? _____
   iv. What is the probability of getting a vegetable for lunch? _____

Evaluation: Let us see how you did.

Learning outcome	No!	Working on it	Yes!
Did you get all the answers?	☹	😐	🙂
Did you get most questions right?	☹	😐	🙂
Did you retry the question(s) you got wrong?	☹	😐	🙂
Were you able to correct your wrong answers?	☹	😐	🙂
If not, did you seek help from others and/or review the chapter?	☹	😐	🙂

Colour the face that shows how you are doing.

ANSWERS

1. "A strong mind is the best treasure! Education is the key, and math is the map—let's unlock success together!"

2. "Numbers never lie, and neither do the clues! Solve math, solve problems, and solve your way to a brighter future!"

3. "Want to open doors to success? Math and learning are the master keys! Let's turn the lock and explore endless

4. "Detectives need sharp minds, and math is the ultimate training! Every equation is a clue leading to new discoveries!"

5. "Education is power, and math is the code! Crack it, and you'll have the secret formula for life's biggest mysteries!"

Chapter 1

> How did you do? Remember to ask for help if you are unsure about anything.

Prior learning

1. Seven thousand, five hundred and sixty-three

2.

Number	Partition	True	False
20	10 + 5 + 2		✓
45	5 + 20 + 10 + 3 + 7	✓	
48	12 + 12 + 12 + 6 + 4 + 2	✓	

3.

a. ½	b. ¼	c. ¾	d. 4/8 or ½
e. $\frac{1}{3}$	f. $\frac{1}{6}$	g. $\frac{4}{6}$ or $\frac{2}{3}$	h. $\frac{6}{10}$ or $\frac{3}{5}$

4.

a. 358	b. 941	c. 113	d. 189	e. 325	

5.

a. 92	b. 282	c. 290	d. 161	e. 201	f. 103

6.

a. 1000	b. 0	c. 2000	d. 3000	

1.1 Practice Questions

Values	1	2	3	4	5	6
Number	26	675	1,398	38,456	203,685	9,754,219
Face value	6	6	9	3	2	9
Place value	1	100	10	10,000	10,000	1,000,000
True value	6	600	90	30,000	200,000	9,000,000

1.2 Practice Questions

a. 87 = 8 x 10 + 7 x 1	f. 370,864 = 3 x 100,000 + 7 x 10,000 + 0 x 1000 + 8 x 100 + 6 x 10 + 4 x 1
b. 296 = 2 x 100 + 9 x 10 + 6 x 1	g. 697,325 = 6 x 100,000 + 9 x 10,000 + 7 x 1000 + 3 x 100 + 2 x 10 + 5 x 1
c. 704 = 7 x 100 + 0 x 10 + 4 x 1	h. 7,598,620 = 3 x 100,000 + 7 x 10,000 + 2x1000 + 9x100 + 7x10 + 8x1
d. 2,978 = 2 x 1000 + 9 x 100 + 7 x 10 + 8 x 1	i. 8,659,423 = 8 x 1,000,000 + 6 x 100,000 + 5 x 10,000 + 9 x 1000 + 4 x 100 + 2 x 10 + 3 x 1
e. 4,043 = 4 x 1000 + 0 x 100 + 4 x 10 + 3 x 1	j. 4,097,351 = 4 x 1,000,000 + 0 x 100,000 + 9 x 10,000 + 7 x 1000 + 3 x 100 + 5 x 10 + 1 x 1

1.3 Practice Questions

1.

	Standard form	Worded form
a.	2,469	two thousand, four hundred and sixty-nine
b.	3,876	three thousand, eight hundred and seventy-six
c.	28,964	twenty-eight thousand, nine hundred and sixty-four
d.	39,453	thirty-nine thousand, four hundred and fifty-three

e.	826,351	eight hundred and twenty-six thousand, three hundred and fifty-one
f.	753,826	seven hundred and fifty-three thousand, eight hundred and twenty-six
g.	1,011,011	one million, eleven thousand and eleven
h.	1,204,507	one million, two hundred and four thousand, five hundred and seven
i.	2,003,014	two million, three thousand and fourteen
j.	9,876,543	nine million, eight hundred and seventy-six thousand, five hundred and forty-three

2.

	Worded form	Expanded form
a.	four hundred and seventy-six	$4 \times 100 + 7 \times 100 + 6 \times 1$
b.	eight thousand, four hundred and twenty-five	$8 \times 1,000 + 4 \times 100 + 2 \times 10 + 5 \times 1$
c.	twenty-six thousand, eight hundred and eight	$2 \times 10,000 + 6 \times 1,000 + 8 \times 100 + 0 \times 10 + 8 \times 1$
d.	two hundred and twenty-eight thousand, six hundred and seventy	$2 \times 100,000 + 2 \times 10,000 + 8 \times 1,000 + 6 \times 100 + 7 \times 10 + 0 \times 1$
e.	eight million, seven hundred and five thousand and two	$8 \times 1,000,000 + 7 \times 100,000 + 0 \times 10,000 + 5 \times 1000 + 0 \times 100 + 0 \times 10 + 2 \times 1$
f.	twenty-four	$2 \times 10 + 4 \times 1$
g.	eight hundred and forty-six	$8 \times 100 + 4 \times 10 + 6 \times 1$
h.	seven thousand, three hundred and sixty-nine	$7 \times 1,000 + 3 \times 100 + 6 \times 10 + 9 \times 1$
i.	eighty-five thousand, eight hundred and sixty-five	$8 \times 10,000 + 5 \times 1,000 + 8 \times 100 + 6 \times 10 + 5 \times 1$
j.	six hundred and two thousand and sixty	$6 \times 100,000 + 0 \times 10,000 + 2 \times 1,000 + 0 \times 100 + 6 \times 10 + 0 \times 1$
k.	eight million, seven hundred and forty-two thousand, eight hundred and sixty-five	$8 \times 1,000,000 + 7 \times 100,000 + 4 \times 10,000 + 2 \times 1,000 + 8 \times 100 + 6 \times 10 + 5 \times 1$

3.

Worded form	Standard form
a. Four hundred and seventy-six	476
b. Eighty thousand, four hundred and twenty-five	8425
c. Twenty-six thousand eight hundred and eight	26,808
d. Two hundred and twenty-eight thousand, six hundred and seventy	228,670
e. Six million, seven hundred and five thousand and two,	6,705,002

4.

Column A	Column B
Buying a car	✓ Buying a bus
✓ Distance from Kingston to Ocho Rios	Distance from your home to your school
✓ The population of Jamaica	Population of Kingston
Distance from Earth to the moon	Distance from Earth to the Sun
The amount of money needed to buy groceries for one week at home	✓ The amount of money needed to buy a school's canteen supplies for one week

Chapter 2

Prior Learning

1. i. Null set, empty set or a set with no members.   ii. The elements of set A are 2, 4 and 6.

    iii. Set A is a set of even numbers between 1 and 7.   iv. Equivalent set

2. Tom has $72

3. Set A = $ 450

2.1-2.4 Practice questions

1. a. Set P contains all letters of the alphabet between l and u. n(P) = 8
    b. Set K contains six types of utensils. n(K) = 6
    c. Set N contains all counting numbers less than 10. n(N) = 9
2. a. Set V is the set of all vowels in the English alphabet.
    b. Set M is the names of five types of clothing.
    c. Set Y is a set of 3 types of furniture.
    d. Set T is the names of 5 types of tools used by carpenters.
    e. Set C contains four farm animals.
3. c. A set is a group of things in a bracket.
4. b. Set R = {2, 3, 6}
5. b. There are no elements in the set.
6. d. The members of the set cannot be counted.
7. n(F) = 7
8. Out of place: electric iron
    Explanations: All the members in set C are stationeries, except the electric iron.

9.

a.	Set T = {1, 2 , 3, 4}	g.	Set Y = {4, 8, 12, 16}
b.	Set C = {5, 6, 7, 8}	h.	Set G = {3, 7, 11, 15}
c.	Set R = {9, 10, 11, 12}	i.	Set S = {1, 5 ,9, 13}
d.	Set P = {13, 14, 15, 16}	j.	Set L = {4, 8, 12, 16}
e	Set X = {1, 5, 9, 13}	k.	Set A = {1, 2, 3, 4, 5, 6, 7, 8, 9, 10, 11, 12, 13, 14, 15, 16}
f.	Set B = {2, 6, 10, 14}	l.	Set P = { }

Chapter 3

Prior learning

1.

a. $\frac{1}{2}$ or one-half	b. $\frac{1}{3}$ or one-third	c. $\frac{3}{4}$ or three-quarters	d. $\frac{4}{5}$ or four-fifths
e. $\frac{3}{6}$ or three sixth	f. $\frac{5}{8}$ or five-eight		

2.

| a. $\frac{1}{3}, \frac{2}{3}, \frac{3}{3}$ | b. $\frac{1}{4}, \frac{2}{4}, \frac{3}{4}, \frac{4}{4}$ | c. $\frac{1}{5}, \frac{2}{5}, \frac{3}{5}, \frac{4}{5}, \frac{5}{5}$ | d. $\frac{1}{7}, \frac{2}{7}, \frac{3}{7}, \frac{4}{7}, \frac{5}{7}, \frac{6}{7}, \frac{7}{7}$ |

3.1 Practice questions

a. $1\frac{1}{2}$	b. $1\frac{4}{5}$	c. $1\frac{7}{8}$
d. $2\frac{2}{3}$	e. $4\frac{3}{4}$	

3.3 Practice questions

a. Any multiply of $\frac{1}{3}$. For example $\frac{2}{6}$ or $\frac{3}{9}$ or $\frac{4}{12}$

b. Any multiply of $\frac{2}{5}$. For example $\frac{4}{10}$ or $\frac{6}{15}$ or $\frac{12}{20}$

c. Any multiply of $\frac{3}{6}$. For example $\frac{1}{2} = \frac{3}{6} = \frac{6}{12}$

d. Any multiply of $\frac{10}{20}$. For example $\frac{5}{10} = \frac{10}{20} = \frac{20}{40}$

3.4 Practice questions

1. From largest to smallest $\frac{25}{30} = \frac{5}{6} > \frac{24}{30} = \frac{4}{5} > \frac{20}{30} = \frac{2}{3}$

2. From smallest to largest

a. $\frac{25}{30} = \frac{1}{2} > \frac{24}{30} = \frac{4}{5} > \frac{20}{30} = \frac{2}{3}$

c. $\frac{9}{12} = \frac{3}{4} > \frac{6}{12} = \frac{1}{2} > \frac{4}{12} = \frac{1}{3}$

b. $\frac{48}{60} = \frac{4}{5} > \frac{45}{60} = \frac{3}{4} > \frac{25}{60} = \frac{2}{3}$

d. $\frac{20}{30} = \frac{2}{3} > \frac{18}{30} = \frac{3}{5} > \frac{15}{30} = \frac{1}{2}$

3.5 Practice questions

1. A = $\frac{3}{5}$	2. $G = \frac{1}{5}$	3. T = $\frac{2}{8} = \frac{1}{4}$	4. N = $\frac{1}{3}$

3.6.1 Practice questions

a. $3\frac{2}{7}$	b. $\frac{3}{5}$	c. $2\frac{2}{3}$	d. $1\frac{3}{8}$	e. $\frac{4}{9}$	f. $3\frac{1}{8}$	g. $3\frac{2}{5}$	h. $\frac{1}{4}$	i. $1\frac{1}{6}$	j. $1\frac{4}{7}$

3.6.2 Practice questions

1.

a. $\frac{1}{2}$	b. $\frac{1}{5}$	c. $1\frac{2}{3}$	d. $1\frac{5}{8}$	e. $1\frac{3}{5}$	f. $\frac{2}{5}$	g. $1\frac{2}{5}$	h. $1\frac{5}{7}$	i. $1\frac{3}{8}$	j. $\frac{1}{8}$

2. a. $4 - 2\frac{1}{3} = 1\frac{2}{3}$ chocolate bars left

   b. $5 \div \frac{1}{2} = 10$ students

   c. $4 \div 8 = \frac{1}{2}$ bun each

d. $3\frac{1}{2} \div 7 =$ Each person gets $\frac{1}{2}$

e. 3 uneven shares is $\frac{1}{8}, \frac{2}{8}, \frac{5}{8}$ or $\frac{1}{8}, \frac{3}{8}, \frac{4}{8}$

Chapter 4

Prior learning

1. a. 66, b. 79, c. 25, d. 27
2. a. 68 pencils, b. $5
3. a. 684 is six hundred and eighty-four = 6 x 100 + 8 x 10 + 4 x 1

   b. 874 and 8 x 100 + 7 x 10 + 4
4. a. 30, b. 80, c. 60, d. 70, e. 70, f. 117, g. 92, h. 62, i. 49, j. 20

4.1.1 Practice questions

a. 10, b. 20, c. 30, d. 50, e. 80, f. 100, g. 100, h. 200, i. 400, j. 700, k. 1800, l. 2900, m. 1000, n. 1000, o. 2000, p. 5000, q. 8000, r. 8000

4.1.2 Practice questions

a. 70; 69, b. 70; 70, c. 90; 87, d. 60; 62, e. 90; 95, f. 40; 42, g. 70; 64, h. 20; 17, i. 50; 55, j. 30; 33

4.2.1 Practice questions

1. a. 10 + 4 + 8 = 8 + 10 + 4 = 4 + 8 + 10 = 22   b. 9 + 7 + 12 = 12 + 9 + 7 = 7 + 12 + 9 = 28

2. a. (12 + 5) + 9 = (12 + 9) + 5 = (5 + 9) + 12 = 26

   17 + 9 = 21 + 5 = 14 + 12 = 26

   b. (9 + 7) + 12 = (12 + 9) + 7 = (7 + 12) + 9 = 28

   16 + 12 = 21 + 7 = 19 + 9 = 28

3. (6 + 11) + 4 = (11+ 4) + 6 = **(6 + 4) + 11** = **21**

4. a. n= 9, b = 16   5. c. P = 5

4.2.2 Practice questions

1. For $a - b \neq b - a$

   $9 - 5 \neq 5 - 9$

   Note: 9 - 5 = 4, but $5 - 9 \neq 4$

2. c. $7 - 8 \neq 1$ 3. a. 8 - 7 = 1   4. a

4.3 Practice question

a. 82, b. 97, c. 83, d. 81, e. 65, f. 54, g. 64, h. 86, i. 95, j. 101, k. 9, l. 5, m. 15, n. 19, o. 18, p. 8, q. 19, r.17, s. 28, t. 23.

CHEETAH
Connect to Higher Education, Electronic Tools, Aplication and Help

Chapter 5

5.1 and 5.2 Practice questions

1. To be completed with a teacher or as a group activity.
2. To be completed with a teacher or as a group activity.
3. Line 1: XX Line 2: XX Line 3: XX Answers to be added when draft version of the book is printed.
4. Length of book: XX Answers to be added when draft version of the book is printed.
5. Length of line: XX Answers to be added when draft version of the book is printed.
6. Line AB: XX Line CD: XX Difference: XX Answers to be added when draft version of the book is printed.

5.3. Practice questions Answers to be added when draft version of the book is printed.

1 a. xx   b. xx   c. xx   d. xx   e. xx

2. 7km

3. a. 400m   b. 600m   c. 800m   d. 1800m

5.4 Practice question

1.

A. 0.1 cm	B. 1.5 cm	C. 2.5 cm	D. 3.5 cm	E. 4.4 cm
F. 5.6 cm	G. 7.2 cm	H. 8.9 cm	I. 9.7 cm	J. 11.8 cm

2.

A. 0.01 cm	B. 0.05 cm	C. 0.12 cm	D. 0.18 cm	E. 0.3 cm
F. 0.45 cm	G. 0.56 cm	H. 0.84 cm	I. 0.88 cm	J. 1.0 cm

3. 0.13 m and 20 cm   4. 0.5 m   5. 0.5 m   6. 2.05 m   7. 1 m

8. 1 m and 10 cm   9. 4 m and 60 cm   10. 8 m

5.5 Practice questions

a. 12:30 or half past 12	b. 12:15 or fifteen minutes past 12	c. 9:15 or fifteen minutes past 9 o'clock	d. ten minutes past 4 in the morning	e. twenty-five minutes to 1 o'clock in the afternoon
f. fifteen minutes to 6 o'clock in the morning	g. five minutes past 12 o'clock	h. twenty minutes past 2 o'clock	i. twenty-nine minutes to 7 o'clock	j. ten minutes 9 o'clock
k. thirteen minutes to 5 o'clock	l. twenty minutes past 10 o'clock	m. twelve minutes past 9 o'clock in the night	n. five minutes to 12 o'clock in the morning	o. twenty minutes past 1 o'clock

CHEETAH™
Connect to Higher Education, Electronic Tools, Aplication and Help

5.6 Practice questions

1. 15 minutes	2. 2 ½ hours	3. 3 minutes	4. 32 minutes	5. No, he is not late. The party starts in 10 minutes
6. 40 minutes	7. 45 minutes	8. 7:15 p.m.	9. 2 hours and 25 minutes	10. 5:52 p.m. or 8 minutes to 6 o'clock

5.7 Practice questions

1. 7 ml	2. 32 ml	3. 15 ml	4. 57 ml	5. 80 ml
6. 110 ml	7. 3 ml	8. 10 ml	9. 1000 ml	10. 26 cm^3 - 18 cm^3 = 8 cm^3

5.8 Practice question

1. B   2. two litres   3. 1 litre and 750 ml

Chapter 6

Prior learning

Prefix	To conversion to g
kilo	divide by 10
deci	divide by 1000 or multiply by $\frac{1}{1000}$
centi	multiply by 1000
milli	divide by 100

Units	Instruments
kg	thermometer
^{0}C	measuring cup
m	scale balance
l	ruler
cm	tape measure

6.1 Practice questions

1. a. grams
   b. kilograms
   c. grams
   d. grams
   e. kilograms
   f. grams
   g. kilogramse

6.2 Practice questions

1.

a. 80 g	b. 89 g	c. 36 g	d. 28 g	e. 45 g
f. 14 g	g. 1 kg	h. 4 kg	i. 6.6 kg	

CHEETAH™
Connect to Higher Education, Electronic Tools, Aplication and Help

2.

a.	2 kg =	2,000 g	i.	17 kg =	17,000 g	m.	9000 g =	9 kg
b.	3 kg =	3,000 g	j.	20 kg =	20,000 g	n.	10,000 g =	10 kg
c.	5 kg =	5.000 g	k.	24 kg =	24,000 g	o.	12000g =	12 kg
d.	8 kg =	8,000 g	h.	1,000 g =	1 kg	p.	15000g =	15 kg
e.	10kg =	10,000 g	i.	2000 g =	2 kg	q.	19000g =	19 kg
f.	11 kg =	11,000 g	j.	4000 g =	4 kg	r.	23,000 g =	23 kg
g.	13 kg =	13,000 g	k.	7000 g =	7 kg			
h.	16 kg =	16.000 g	l.	8000 g =	8 kg			

3.

a.	1 kg	150 g=	1,150 g	i.	18 kg	974 g=	18,974 g	q.	5,000 g =	5 kg	0 g
b.	2 kg	200 g=	2,200 g	j.	21 kg	356 g=	21,356 g	r.	6,439 g =	6 kg	435 g
c.	4 kg	380 g=	4,380 g	k.	24 kg	778 g=	24,778 g	s.	7,103 g =	7 kg	103 g
d.	6 kg	500 g=	6,500 g	l.	1,200 g =	1 kg	200 g	t.	8,002 g =	8 kg	2g
e.	7 kg	250 g=	7,250 g	m.	1,000 g =	1 kg	0 g	u.	9,134 g =	9 kg	134 g
f.	13 kg	600 g=	13,600 g	n.	1,234 g =	1 kg	234 g	v	10,001 g =	10 kg	1 g
g.	15 kg	151g=	15,151 g	o.	2,500 g =	2 kg	500 g				
h.	16 kg	234 g=	16,234 g	p.	3,550 g =	3 kg	550 g				

4.

a. 28 g   b. 74 g   c. 2 kg   d. 17 g   e. 26 g   f. 9 kg

6.3 Practice questions
1. C   2. C   3. 4,000 kg
4. No. 6 tonnes is over the maximum weight of the vehicles allowed to cross the bridge. It is unsafe to cross the bridge with a vehicle that is too heavy.
5. 8,500 kg

6.4 Practice questions
1. b   2. d   3. b   4. b   5. a   6. d

6.5 Practice questions
1. 26 ^{0}C   2. 46 ^{0}C   3. 60 ^{0}C   4. 65 ^{0}C   5. 50 ^{0}C   6. 6 ^{0}C

6.6 Practice questions
1. A = 2 ^{0}C, B = 17 ^{0}C, Difference = 15 ^{0}C, A is colder
2. A = 56 ^{0}C, B= 89 ^{0}C, Difference = 33 ^{0}C, B is warmer
3. A = 2 ^{0}C, B= 13 ^{0}C, Difference = 11 ^{0}C, A is colder
4. A = 27 ^{0}C, B= 9 ^{0}C, Difference = 18 ^{0}C, A is warmer
5. A = 55 ^{0}C, B = 48 ^{0}C, Difference = 7 ^{0}C, B is colder
6. P = 30 ^{0}C, Q = 20 ^{0}C, Difference = 10 ^{0}C, P is hotter
7. The office became warmer. Temperature difference = 7 ^{0}C
8. The workers put their sweaters on because the office became much colder.
9. Warmer. In the Caribbean, the temperature outside is usually greater than 25 ^{0}C. By opening the window, she let the cold air go outside and the warm air come inside the office.
10. 31 ^{0}C

Chapter 7

Prior learning

Measurement	Base units	Units of measurement
Mass (weight)	__grams__	__g__ or kg
__Length__	metre	__m__ or cm
Volume or capacity	__litre__	__l__
__Time__	Seconds, _minutes_ or _hours_	__s__ mins or hr.
__Temperature__	degree Celsius	__°C__

7.1 and 7.2 Practice questions

Prefixes	Relationship between prefixes and base unit	
deka	__10__	times larger than base unit
hecto	__100__	times larger than base unit
kilo	__1000__	times larger than base unit
milli	__1000__	times smaller than base unit
centi	__100__	times smaller than base unit
deci	__10__	times smaller than base unit

1. **Converting metric units for lengths and distances**

	Cm to mm		mm to cm	
i.	8 cm to mm	80 mm	xvi. 20 mm to cm	2 cm
ii.	10cm to mm	100 mm	xvii. 60 mm to cm	6 cm
iii.	46 cm to mm	460 mm	xviii. 120 mm to cm	12 cm
iv.	351 cm to mm	3510 mm	xiv. 200 mm to cm	20 cm
v.	728 cm to mm	7280 mm	xx. 980 mm to cm	98 cm
vi.	2 m to cm	200 cm	xxi. 100 cm to m	1 m
vii.	5 m to cm	500 cm	xxii. 300 cm to m	3 m
viii.	26 m to cm	2600 cm	xxiii. 700 cm to m	7 m
ix.	36 m to cm	3600 cm	xxiv. 1000 cm to m	10 m
x.	136 m to cm	13,600 cm	xxv. 1,200 cm to m	12 m
xi.	1 km to m	1000 m	xxvi. 2000 m to km	2 km
xii.	2 km to m	2000 m	xxvii. 3000 m to km	3 km
xiii.	3 km to m	3000 m	xxviii. 5000 m to km	5 km
xiv.	6 km to m	6000 m	xxiv. 2000 m to km	2 km
xv.	9 km to m	9000 m	xxx. 56,000 m to km	56 km

2. $3\frac{1}{2}$ metres.

3. 600 + 400 + 600 + 400 = 2000 cm

In metre $= \frac{2000}{100} = 20$ m

CHEETAH™
Connect to Higher Education, Electronic Tools, Aplication and Help

4.

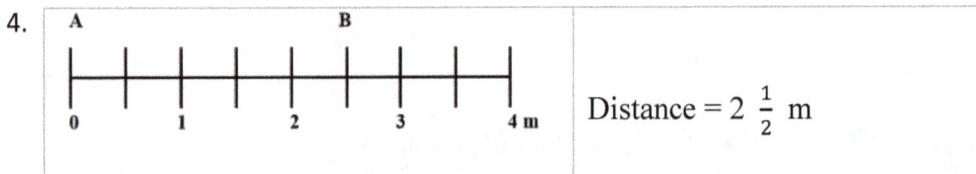

Distance = $2\frac{1}{2}$ m

5. 1 m = 100 cm. The book is a rectangle. There are 2 long sides and 2 short sides.

The length of the two long sides is 30 cm + 30 cm = 60 cm. So, the length remaining for the

two short sides is 100 cm – 60 cm = 40 cm

Therefore, the length of one of the short side is $\frac{40\ cm}{2}$ = 20 cm.

6. 10 x 50 cm = 500 cm = 5 m   7. 2 m + 1 ½ m = 3 ½ m = 350 cm

8. 6 cm + 8 cm = 14 cm = 140 mm   9. 2 m + 52 m = 54 m = 5,400 cm

10. 3 m = 300 cm

11. All five cars occupy 2 ½ m + 2 ½ m + 2 ½ m + 2 ½ m + 2 ½ m = 12 ½ m of space

The space in between the cars is ½ m + ½ m + ½ m + ½ m = 2 m

All the cars and space between occupy 12 ½ m + 2 m = $14\frac{1}{2}$ of space.

7.4 Practice Questions

1.

a. 2,000 ml = 2 L	d. 1,500 ml = 1 ½ L	g. 8,000 ml = 8 L
b. 15,000 ml = 15 L	e. 1,000 ml = 1 L	h. 25,000 ml = 25 L
c. 8,500 ml = 8.5 L	f. 6,000 ml = 6 L	i. 10,500 ml = 10 ½ L

2.

a. 2 L = 2,000 ml	d. 3 ½ L = 3,500 ml	h. 7 ½ L = 7,500 ml
b. 7 L = 7,000 ml	e. 8 L = 8,000 ml	i. 8 ½ L = 8,500 ml
c. 10 L = 10,000 ml	f. 50 L = 50,000 ml	k. 120 L = 120,000 ml

3.

a. 1 ½ L = 1L and 500 ml	c. 5 ½ L = 5 L and 500 ml	e. 8 ½ L = 8 L and 500 ml
b. 2 ½ L = 2 L and 500 ml	d. 7 L = 7 L and 0 ml	f. 10 ½ L =10 L and 500 ml

4.

a. 2355 ml = 2 L and 355ml	c. 7500 ml = 7 L and 500 ml	e. 8364 ml = 8 L and 364 m
b. 6001 ml = 6 L and 1 ml	d. 5356 ml = 5 L and 356 ml	f. 38545 ml = 38 L and 545 ml

5. 2 L = 2,000 ml. If 20 ml leaks out in 1 minute, then 20ml x 10 = 200 ml leaks out in 10 minutes. Amount left in the pan after 10 minutes is 2,000 ml – 200 ml = 1,800 ml.

7.5 Practice Questions

1.

a. 1 g = 1,000 mg	d. 2 ½ g = 2,500 mg	g. 5 ½ g =5,500 mg
b. 6 g = 6,000 mg	e.7 g = 7,000 mg	h. 8 ½ g = 8,500 mg
c. 16 g = 16,000 mg	f. 25 g = 25,000 mg	i. 220 g = 220,000 mg

2.

a. 1,000 mg = 1 g	d. 2,000 mg = 2 g	g. 5,000 mg = 5 g
b.7, 000 mg = 7 g	e. 9,500 mg = 9 ½ g	h.15,000 mg = 15 g
c.1,500 mg = 1 ½ g	f. 3,500 mg = 3 ½ g	i. 4,500 mg = 4 ½ g

3.

a. 1,000 g = 1 kg	d. 3,000 g = 3 kg	g. 6,000 g = 6 kg
b. 7, 000 g = 7 kg	e. 8,500 g = 8 ½ kg	h. 12,000 g = 12 kg
c. 1,500 g = 1 ½ kg	f. 2,500 g = 2 ½ kg	e. 5,500 g = 5 ½ kg

4.

a. 2 ½ g = 2 g and 500 mg	d. 3 ½ g = 3 g and 500 mg	g. 8 ½ g = 8 g and 500 mg
b. 13 ½ g =13 g and 500 mg	e. 6 ½ g = 6 g and 500 mg	h. 17 ½ g = 17 g and 500 mg
c. 260 mg = 0g and 260 mg	f. 4900 mg = 4 g and 900 mg	i. 7500 mg = 7 g and 500 mg

5.

a. 3,401 g = 3 kg and 401 g	c. 4,600 g = 4 kg and 600 g	e. 8,294 g = 8 kg and 294 g
b. 5,755 g = 5 kg and 755 g	d. 5,660 g = 5 kg and 660 g	f. 10,545 g = 10 kg and 545g

6.

a. 3kg and 1g = __3,001__ g	d. 5 kg and 6 g = __5,006__ g
b. 6 kg and 755 g = __6,755__ g	e. 14 kg and 609 g = __14,609__ g
c. 7 kg and 294 g = __7,294__ g	f. 8 kg and 55 g = __8,055__ g

7. 3,500 g

8. 12 ½ kg = 12,500 g. So, the load weighs 12,500 g – 500 g = 12,000 g or 12 kg

9. 1 kg = 1,000 g. So, 5 ½ kg = 5,500 g

10. Mass of 3 student lunches is 1 ½ kg + 2 kg + 2 kg = 6 kg

Mass of the 4th student's lunch is 7 ½ kg – 6 kg = 1 ½ kg

Mass of the 4th student's lunch in grams is 1 ½ kg x 1000 = 1,500 g

CHEETAH
Connect to Higher Education, Electronic Tools, Aplication and Help

7.6 and 7.7 Practice questions

1.

a. 60 sec =	1 mins.	i. 400 secs =	6 mins and 40 secs	q. 720 mins =	12 hrs.
b. 120 sec =	2 mins.	j. 500 secs =	8 mins and 20 secs	r. 1800 mins =	30 hrs.
c. 180 sec =	3 mins.	k. 600 secs =	10 mins and 0 secs	s. 2400 mins =	40 hrs.
d. 240 sec =	4 mins.	l. 1220 secs =	20 mins and 20 secs	t. 245 mins =	4 hrs. and 5 mins
e. 480 sec =	8 mins.	m. 60 mins =	1hrs.	u. 400 mins =	6 hrs. and 40 mins
f. 600 sec =	10 mins.	n. 240 mins =	4 hrs.	v. 500 mins =	8 hrs. and 20 mins
g. 1200 sec =	20 mins.	o. 360 mins =	6 hrs.	w. 560 mins =	9 hrs. and 20 mins
h. 365 secs =	6 mins and 5 secs	p. 420 mins =	7 hrs.	x. 720 mins =	12 hrs. and 0 mins

2.

300 second	135 seconds
2 minutes and 15 second	120 minutes
2 hours	14 days
72 hours	5 minutes
2 weeks	3 days

3. 150 mins   4. 5 hours   5. 2 mins   6. 240 mins   7. 7 mins   8. 210 mins
9. 480 mins   10. 360 mins

Chapter 8

Prior learning

Column A	Column B
Point	Point where two-line segments meet at a right angle (90 °C).
Line segment	A location in space, usually represented by a dot.
Simple closed path	A straight path between two points.
Square corner	A line that starts and ends at the same point.

8.1 and 8.2 Practice questions

1. C   2. D
3. B   4. D
5. A
6. B   7. C
8. D   9. D
10. B

8.3 Practice questions

1. C

2.

vi. a.	vii. b.
A / O / B — < AOB is an acute angle	C / O / D — < COD is an acute angle
viii. c.	ix. d.
P / O / Q — < POQ is a right angle	M N / O L — <MOL is obtuse / < MON and < NOL are acute

3. i

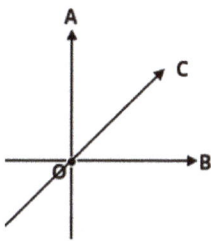

ii. = C

4. B   5. A   6. C

8.4 Practice question

1. B   2. A   3. B   4. C   5. B
6. 90^0 7. 90^0 8. b = 45^0 9. D   10. D

8.5 Practice question

1. a. $\overrightarrow{AOC}$ = f   b. $\overrightarrow{COD}$ = g   c. $\overrightarrow{DOE}$ = h

2. Use any Capital or common letter.   3. p = AOB,   q = BOC

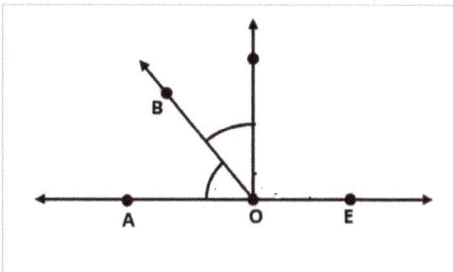

8.6. Practice questions

1.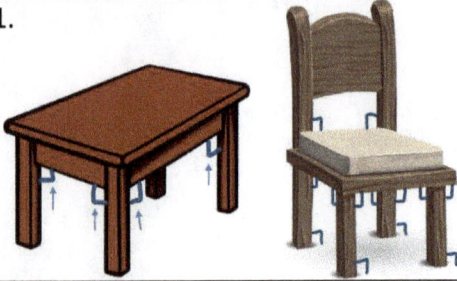

2.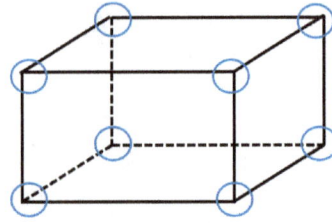

3. Each sheet of paper has 4 square corners and there are 9 sheets. 4 x 9 = 36

4.
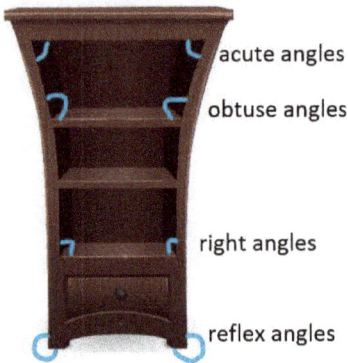

acute angles

obtuse angles

right angles

reflex angles

8.7 Practice question

1.
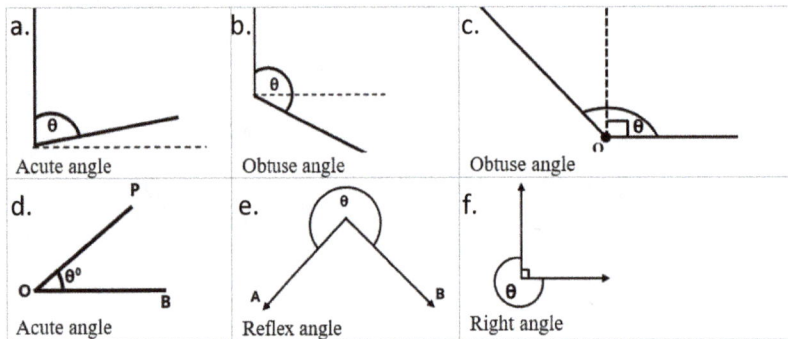

a.	b.	c.
Acute angle	Obtuse angle	Obtuse angle
d.	e.	f.
Acute angle	Reflex angle	Right angle

Note: Ө - theta

2. Acute – BCO, BOC   Obtuse- ABC, BCD   Right angle – AOB, ODC

8.8 Practice questions

1. b   2. d   3. c   4. d   5. a
6. b   7. a   8. b   9. c

CHEETAH
Connect to Higher Education, Electronic Tools, Aplication and Help

Chapter 9

Prior learning

1. B

2.

kite   rectangle   triangle   square

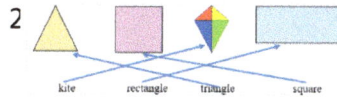

9.1 Practice questions

1. Congruent – all sides have the same lengths and angles.

2a. Congruent – triangle will fit exactly over one another.

2b. Not congruent – triangle will not fit exactly, over each other. The angle between the two equal

   sides are not equal.

3a. Figures are not congruent. They do not fit exactly over each other.

3b. When turned around they fit exactly over each other.

4. No triangle are not congruent. They do not fit exactly over each other.

5. Figures are congruent, each covers the same space and would fit over the other.

6. The pages of a book are congruent.

7. Figures is same size fit over each other.

8. A + B are not congruent. They are not the same size and cannot fit over each other.

9. Both are congruent. Just turn them over and they fit over each other.

10. Yes, the dotted line divides the square exactly in two. Each angle fits over the other.

9.2 Practice questions

1.

	Shape	Dimensions	Images
a.	equilateral triangle	3 sides, 3 angles, 3vertices	
b.	square	4 sides, 4 angles, 4 vertices	
c.	pentagon	5 sides, 5 angles, 5 vertices	
d.	hexagon	6 sides, 6 angles, 6 vertices	
e.	Heptagon	7 sides, 7 angles, 7 vertices	
f.	Octagon	8 sides, 8 angles, 8 vertices	

2. A and D   3. 6 figures

4. Regular: ABKJ, BCDK, KDFG, JKGH, ACFH
   Irregular: ABGH, BCFG, ACDJ, JDFH

5. Pentagon and hexagon

6. a. ABC or EDF   b. AEDC,
   c. FDCBE, ABEDC   d. ABEFDC

7. 2 pairs

9.3 Practice questions

1. 5 triangles   2. 5 squares

3. 2 triangles: Place one triangle on either side of the square.

4.
   6 triangles

5. 2 triangles and 1 square

6. 4 triangles

7. 6 polygons

8. 6 triangles: AFG, ABG, ABF, HED, HDC and DEC.
   2 rectangles: EFGH, BCHG
   2 pentagons: BCDEF, ABCEF
   1 hexagon: ABCDEF

CHEETAH
Connect to Higher Education, Electronic Tools, Aplication and Help

g.	Nonagon	9 sides, 9 angles, 9 vertices	
h.	Decagon	10 sides, 10 angles, 10 vertices	

9.4 Practice questions

a. Triangle- regular shape

b. Triangle - irregular shape

c. Square - regular shape

d. Rectangle - regular shape

e. Right-angled triangle - irregular shape

f. Kite - irregular shape

g. Hexagon - regular shape

h. Pentagon – irregular shape

9.5 Practice questions

Q 1 to 3. Teacher to supervise activity.

Chapter 10

Prior learning

Favourite fruits	Tally	Total
apples	̶H̶H̶ ‖‖	9
plum	̶H̶H̶ ̶H̶H̶	10
peas	‖	2
naseberry	̶H̶H̶‖	6

10.1 Practice questions

1. Annakay's actiona was not correct. A sample is a small part of the total population.

2. Population is all the students at the school. The sample is the 'some students' who were asked about their preferences.

3. Students population is 56 + 85 = 141

4. School population is 8 x 45 = 360

5. Population of grade six students is 304 + 85 = 389

6. Population = 18; sample = 5

7. 140 – (42 + 26) = 72 attendees for the guests population

10.2 Practice question

1. B

2. C

3. D

4. 30 pens

5. 5 rolls

6. 7 pencils

7. 15 bottles

8. 11 ear plugs

9. 10 match sticks

10. 7 boxes

10.3 Practice question

1.

Population	Sample
80	_8_
_60_	6
140	_14_
_120_	12
320	_32_

2.

	Population	Sample amount taken	Good sample	Not good sample
i.	200	16		✓
ii.	150	12		✓
iii.	90	10	✓	
iv.	120	12	✓	
v.	240	23		✓

10.4 Practice question

1. Population = 75; If there are 35 women, then there are 75 - 35 = 40 men.

   Sample of men is $\frac{10}{100}$ x 40 = 4 men

2. Population = 95; If there are 45 men, then there are 95 - 45 = 50 women.

   Sample of women is $\frac{10}{100}$ x 50 = 5 women

3. Sample peas to be plant is $\frac{10}{100}$ x 80 = 8 peas

4. No. A 12 men sample is less than the minimum 10% for choosing a suitable sample.

5. Yes. The sample was suitable because it exceeded the minimum 10% for choosing a suitable sample.

6. Sample = $\frac{10}{100}$ x 30 = 3 students

7. Sample = $\frac{10}{100}$ x 300 = 30 students

8. Suitable sample = $\frac{10}{100}$ x 80 = 8 boy

9. Suitable sample = $\frac{10}{100}$ x 120 = 12 girls

10. C

Chapter 11

11.1 Practice questions

1. & 2. Answers for the tally tables will vary depending on the number of students in your class and each student's response to the question.

3.

Chicken lunch	Tally	Number
Fried chicken and rice	‖‖‖ ‖‖‖ ‖‖‖‖	14
Stew chicken and rice	‖‖‖ ‖‖‖ ‖‖‖ ‖‖	17
Curried chicken and rice	‖‖‖ ‖‖‖ ‖‖‖ ‖‖‖ ‖‖	22
Fried chicken and chips	‖‖‖ ‖‖‖ ‖‖‖ ‖‖‖ ‖‖‖ ‖‖‖	30
Chicken soup	‖‖‖ ‖‖‖ ‖‖‖ ‖‖‖	18
	Total =	102

4., 5. & 6.
Answers for the tally tables will vary depending on the number of students in your class and each student's response to the question.

11.2 Practice questions

1. Tally table showing phone calls for last week of the month

Days	Tally	Number of calls per day
Sunday	‖‖‖ ‖	6
Monday	‖‖‖ ‖‖‖	8
Tuesday	‖‖‖ ‖‖‖‖	9
Wednesday	‖‖‖ ‖‖‖ ‖	11

Thursday	‖‖‖ ‖‖‖ ‖‖‖	13
Friday	‖‖‖ ‖‖‖‖	9
Saturday	‖‖‖ ‖‖‖ ‖‖‖ ‖‖‖	18

2i). Table showing result of votes student cast for each subject.

Subject	Tally	Number of votes students cast for each subject
Science	‖‖‖‖	4
Drama	‖‖‖ ‖‖	7
Music	‖‖‖ ‖‖	7
Phonics	‖‖‖ ‖‖‖ ‖‖‖‖	14
Art	‖‖‖ ‖‖‖‖	9
Mathematics	‖‖‖ ‖‖‖ ‖‖‖ ‖‖‖	18
Guidance	‖‖‖ ‖‖‖ ‖‖‖ ‖‖‖‖	19
Language Arts	‖‖	2
	Total =	78

ii).

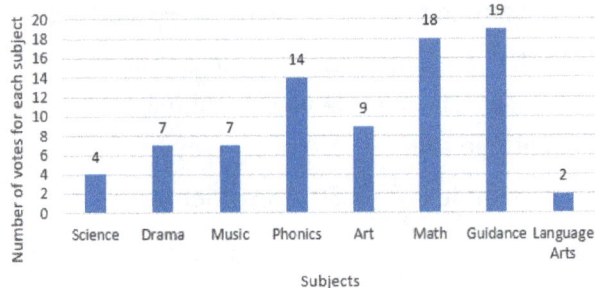

Bar chart showing number of votes cast for each subject

iii). Most students enjoy Guidance; only 2 students enjoyed Language arts.

iv). Graph and bar chart shows how the students loved the subjects in this order (most favored to least favored): Guidance, Math, Phonics, Art, Music, Drama, Science, Language Arts.

3.i). Table on data collected for survey on the size shoes that the classmates wore given.

Shoe size	Tally	Number of shoes
6	‖‖‖‖	4
7	‖‖‖ ‖	6
8	‖‖‖ ‖‖‖	8
9	‖‖‖ ‖‖	7
10	‖‖‖	5
11	‖	1
	Total =	31

ii).

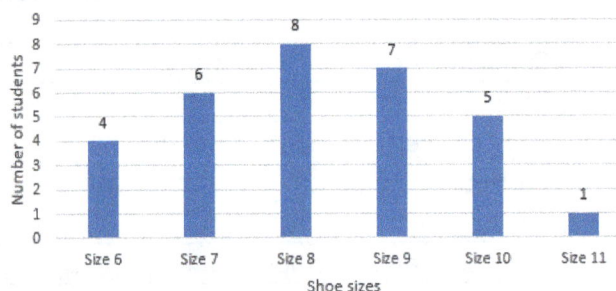

Bar chart showing number of students with different shoes sizes

CHEETAH
Connect to Higher Education, Electronic Tools, Aplication and Help

4.

Bar chart showing favourite colours among 40 students

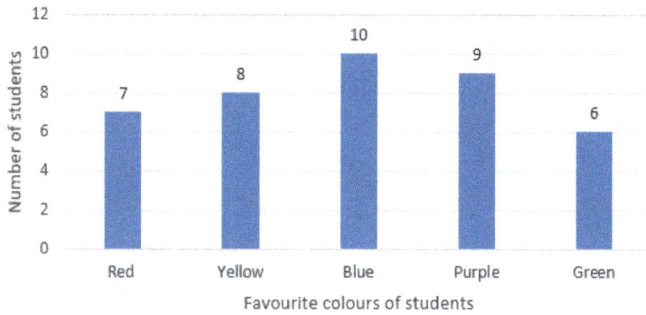

5. Table showing vegetables bought at a farmers' market.

Types of vegetables	Number of vegetable plants
Cabbage	2
Lettuce	4
Spinach	6
Broccoli	8
Callaloo	10
Total =	30

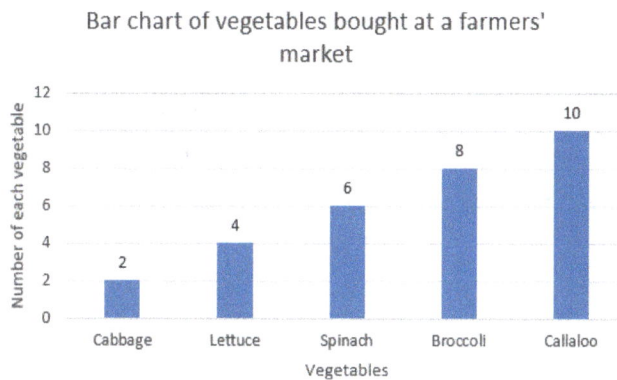

Bar chart of vegetables bought at a farmers' market

6. Same or draw different x axis scale.

Types of books	Number of library books borrowed
Comic books	7
Magazine	5
School textbooks	2
Storybooks	10
Journals	1
Total =	25 books

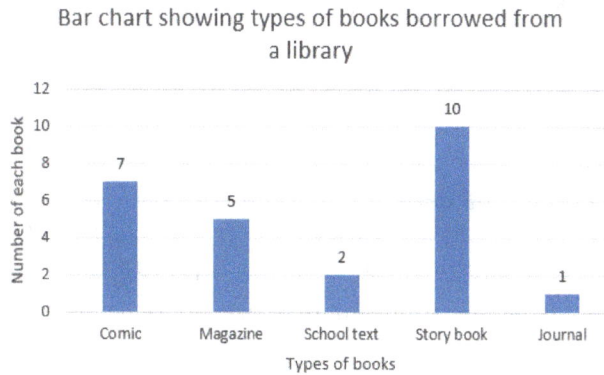

Bar chart showing types of books borrowed from a library

7. Table showing the number of students who preferred each type of movie.

Types of movies	Number of youths preferring each type of movie
Comedy	10
Drama	5
Romance	8
Horror	3
Adventure	8
Total	35

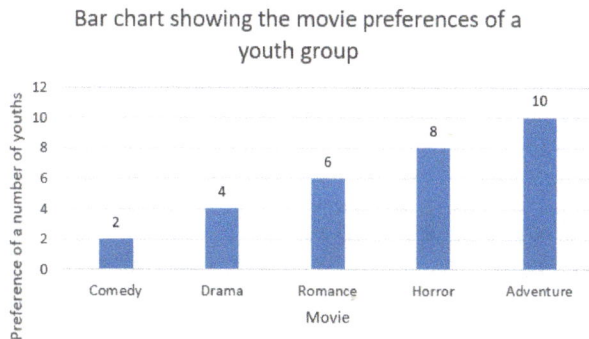

Bar chart showing the movie preferences of a youth group

8. Shops 3 and 4

9. i). Tally table showing the number of hours students spend on social media daily.

Time spent daily of social media (hrs)	Tally of students	Number of students
1	ЖНТ ЖНТ	10
2	ЖНТ II	7
3	ЖНТ I	6
4	III	3
5	II	2

ii). Bar chart showing hours students spent on social media daily

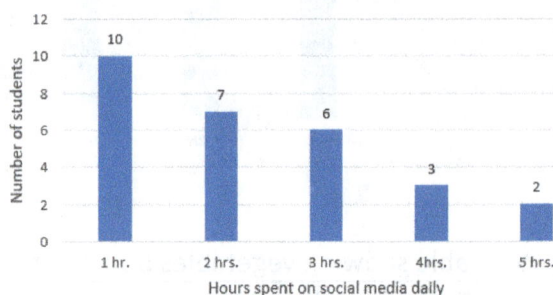

iii). All the students doing the survey spent at least 1 hour on social media daily. Twenty-three (23) students 3 hours or less on social media daily. Five (5) students spent at least 4 hours on social media daily.

Chapter 12

Prior learning

1.

Times	1	2	3	4	5	6	7	8	9	10	11	12
2	2	4	6	8	10	12	14	16	18	20	22	24
3	3	6	9	12	15	18	21	24	27	30	33	36
4	4	8	12	16	20	24	28	32	36	40	44	48

2a. 2 x 4 = 8;   4 x 2 = 8   2b. 3 x 4 = 12; 4 x 3 = 12

3. a. 768 x 2 = 1536   b. 576 x 3 = 1728   c. 896 x 4 = 3584   d. 976 x 5 = 4880

4. Using commutative property, select the true statements.

   a. True   b. True   c. False   d. False   e. True

5. a. A = 4 x 3 = 12; B = 5 X 5 = 25   b. 5 x 6 = 30   c. 6 X 4 = 24 Students

12.1 Practice question

1. a) 2,7697   b) 37,980   c) 57,168   d) 26,887   e) 21,272   f) 51,426

2. a) 619,262   b) 255,864   c) 130,530   d) 271,524   e) 124,257   f) 353,000

12.2 Practice questions

1) 276   2) 765   3) 348   4) 408   5) 686   6) 632   7) 378   8) 375   9) 544   10) 768

12.3 Practice questions

1) 10   2) 80   3) 1,400   4) 8,500   5) 39,000   6) 84,000   7) 48,000   8) 3,560   9) 234,000   10) 16,340

12.4 Practice questions

1) 4,032   2) 2,304   3) 1,945   4) 7,168   5) 4,824   6) 1,020   7) 55,176   8) 197,714   9) 207,582

12.5 Practice questions

Time	1	2	3	4	5	6	7	8	9	10	11	12
1	1	2	3	4	5	6	7	8	9	10	11	12
2	2	4	6	8	10	12	14	16	18	20	22	24
3	3	6	9	12	15	18	21	24	27	30	33	36
4	4	8	12	16	20	24	28	32	36	40	44	48
5	5	10	15	20	25	30	35	40	45	50	55	60
6	6	12	18	24	30	36	42	48	54	60	66	72
7	7	14	21	28	35	42	49	56	63	70	77	84
8	8	16	24	32	40	48	56	64	72	80	88	96
9	9	18	27	36	45	54	63	72	81	90	99	108
10	10	20	30	40	50	60	70	80	90	100	110	120
11	11	22	33	44	55	66	77	88	99	110	121	132
12	12	24	36	48	60	72	84	96	108	120	132	144

12.6 Practice question

1. Find the sum 12 + 20 = 32 Find difference 20 − 12 = 8 Take difference for sum 32 − 8 = 24	2. Find sum 20 + 4 = 24 Find difference 20 − 4 = 16 Add sum and difference 24 + 16 = 40	3. Find product = 12 x 8 = 96 Find sum = 12 + 8 = 20 Add sum to product 96 + 20 = 116
4. Find product 5 + 4 = 9 Find sum = 5 + 4 = 9 Take sum for product 20 − 9 = 11	5. Find difference 12 − 8 = 4 Find product 8 x 12 = 96 Add difference to product 4 + 96 = 100	6. Find difference 30 − 20= 100 Product = 20 x 30 = 600 Divide product by difference $\frac{600}{10}$ = 60
7. Product = 10 x 20 = 200 Difference = 20 − 10 = 10 Divide product by difference $\frac{200}{10}$ = 20	8. Product = 5 x 20 = 100 Sum = 20 + 5 = 25 Divide product by sum $\frac{100}{25}$ = 4	9. Find sum = 20 + 40 = 60 Difference = 40 − 20 = 20 Divide = $\frac{60}{20}$ = 3
10. Allana has 48 pencils If Allana has 4 times as many as John. Then John has = $\frac{48}{4}$ = 12pencils	11. Kimberly has 12 pens. Shawn has 6 times as many as Kimberly. Shawn has 6 x 12 = 72 pens	12. Thadius = 8 marbles Jevaugh has 2 times as many as Thadius = 2 x 8 = 16 marbles Ackeem has 3 times as many as Jevaughn 3 x 16 = 48 marbles Altogether = 8 + 16 + 48 = 72 marbles

CHEETAH
Connect to Higher Education, Electronic Tools, Aplication and Help

12.7 Practice question

1.

Relevant information	Irrelevant Information
2022 school had 840; 2023 school 785	2021 school had 736
Students who left 840 − 785 = 55	

2. Deal with only puppies 6 + 3 = 9 puppies

Relevant information	Irrelevant Information
He save $60 on Monday and $35 on Tuesday and Wednesday	A student needed to save $150 for the week
In all = $60 + $35 + $35 = $130	

3.

Relevant information	Irrelevant Information
Mikey has 6 puppies; His sister has 3 puppies	10 Kittens; His sister gave 6 kittens to the neighbours
Total = 6 + 3 = 9	

4. Elaine read 2 x 8 = 16   5. Janna has = 20 -18 = 2 stones

6. Only length 4 cm and width 3 cm. Perimeter = 4 + 4 + 3 + 3 = 14 cm

12.8 Practice question 6

1. Step 1: Total spent on food = $850 + 900 = $ 1750   Step 2: $2,000 - $1,750 = $250

2. Step 1: Cost of buying 5 books = 5 x 120 = $600; Cost of selling 5 books = 5 x 125 = $625

   Step 2: Money made = $625 − $600 = $25

3. Step 1: Total String length = 45 m + 25 m = 70 m long   Step 2: 100 m − 70 m = 30m left on the roll.

4. Step 1: She paid 5 x $ 500 = $2,500 and 15 x $ 400 = $6,000

   Step 2: She must pay $2500 + $ 6,000 = $8,500

5. If Pen = $50. Pencil = $30 and book = $80   b. 3 ($50) + 3 ($30) + 2 ($80) = $150 + $90 + $160 = $400

6. Step 1. Cost 7 books before discount is $120 x 7 = $840; $20 taken from each cost is $20 x 7 = $140

   Step 2: Final cost of each book is $840 - $140 = $700

Chapter 13

Prior learning

1.

Numerals	0.25	$1\frac{1}{2}$	$\frac{8}{3}$	$\frac{3}{4}$
Proper fraction				✓
Improper fraction			✓	
Mixed fraction		✓		
Decimal fraction	✓			

2. 2. $\frac{1}{2}$ and $\frac{4}{8}$

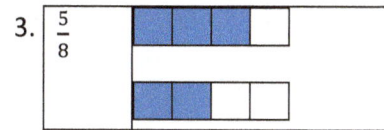

3. $\frac{5}{8}$

4. a. $\frac{7}{5}$ b. $\frac{13}{12}$ c. $\frac{2}{4}$ d. $\frac{4}{8}$

5.
a. $\frac{3}{5}$ + $\frac{2}{5}$ = 1

b. $\frac{2}{4}$ - $\frac{1}{4}$ = $\frac{1}{4}$

6. a. $4\frac{1}{4}$ b. $2\frac{1}{2}$ c. $\frac{5}{10}$ d. $1\frac{5}{8}$

13.1 Practice questions

1. a. 0.1   b. 0.01   c. 0.11   d. 0.16   e. 0.7   f. 0.8   g. 0.55   h. 0.099   i. 0.09   j. 0.03

2. a. $\frac{2}{10} = \frac{1}{5}$ b. $\frac{68}{100} = \frac{17}{25}$ c. $\frac{4}{10} = \frac{2}{5}$ d. $\frac{79}{100}$ e. $\frac{25}{1000} = \frac{1}{40}$ f. $\frac{45}{100} = \frac{9}{20}$ g. $\frac{8}{10} = \frac{4}{5}$ h. $\frac{75}{100} = \frac{3}{4}$ i. $\frac{601}{1000}$

j. $\frac{125}{1000} = \frac{1}{8}$

13.2 Practice question

1. a. $1.45   b. $3.40   c. $2.35   d. $1.75   e. $0.65   f. $8.45   g. $5.30   h. $0.95   i. $7.50   j. $12.34

2. a. 350 ¢   b. 50 ¢   c. 236 ¢   d. 550 ¢   e. 490 ¢   f. 1230 ¢   g. 802 ¢   h. 695 ¢   i. 742 ¢   j. 2531 ¢

13.3 Practice question

1.   a. $31.76   b. 5.731   c. 31.15g   d. $38.99   e. 19.68 m
     f. 36.821   g. 65 ºC   h. 81.855   i. 67.774

2. a. $ 8.89   b. $1.78 = 178 ¢   c. $ 0.52 ¢   d. 8.27   e. $ 9.07   f. $ 0.69 ¢
   g. 8.89   h. $11.88   i. $ 8.47   j. $ 0.989   k. $17.103
   l. $ 1.00 = 100 ¢

3. a. 16.66   b. 13.34   c. 3.936   d. 0.284   e. 55.575
   f. 745.764

g. $2131.35   h. 0.612   i. 75.924   j. $ 257.80

4. a. 32   b. 30.7   c. 30.5   d. 81   e. 57   f. 27.4   g. 31   h. 10.1   i. 30.7

5. a. $63.95 + $38.68 = $ 102.63   b. $138.67 + $149.36 = $288.03
   c. $154.67 + $168.53 = $323.20   d. $0.76 + $0.76 + $0.76 + $0.76 + $0.76 = $3.80
   e. $360.84 + $140.75 + $140.75 = $642.34   f. $294.32 - $175.94 = $118.38
   g. $565.80 - $487.95= $77.85   h. $916 - $568 = $348
   i. $867.90 - $678.50 = $189.40   j. $865.86 - $678.95 = $186.91

6. a. $455.16 x 8 = $3,641.28   b. $556.50 x 18 = $10,017   c. $248.13 x 8 = $1,985.04
   d. $235.50 x 9 =$2,119.50   e. $150.34 x 12 = $1.804.08   f. $3,48.75 ÷ 6 = $58.125
   g. $985.50 ÷ 9 = $109.50   h. $8,645 ÷ 7 = $1,235   i. $12,782.88 ÷ 12 = $1,065.24
   j. $1,320 ÷ $120 per hour = 11 hours

13.4 Practice question

1. $4 \times 10 + 6 \times 1 + \frac{3}{10} + \frac{5}{100}$ 2. $1 \times 100 + 2 \times 10 + 6 \times 1 + \frac{8}{10} + \frac{5}{100}$ 3. $1 \times 1 + \frac{9}{10}$

4. $8 \times 100 + 9 \times 10 + 5 \times 1 + \frac{3}{10} + \frac{2}{100}$ 5. $\frac{8}{10} + \frac{6}{100}$

13.5 Practice question

1. Total collect: $45.50 + $16.84 + $21.65 + $63.85 = $147.84
   Left to be collected: $150.00 - $147.84 = $2.16
2. Total cost: 2 x $124.45 + $450 = $698.90; Change = $750 - $698.9 = $51.10
3. Cost of 3 kg of corn = $85.50 x 3= $256.50
4. Amount spent is $135.67 + $457.85 = $587.51; Amount left is $787.00 - $587.51 = $199.49
5. Amount removed is $126.34 + $346.87 = $473.21; Amount left = $967.85 - $473.21 = $494.64

13.6 Practice question

| a. $2\frac{1}{4}$ | b. $3\frac{5}{6}$ | c. $2\frac{5}{8}$ | d. $3\frac{2}{3}$ |

13.7 Practice question

1. $\frac{1}{4} + \frac{1}{4} = \frac{1}{2}$ day   2. $\frac{3}{8} + \frac{1}{8} = \frac{4}{8}$ or $\frac{1}{2}$ of cake eaten

3. $1 - \frac{3}{5} = \frac{5}{5} - \frac{3}{5} = \frac{2}{5}$ was not paid   4. $1 - \frac{5}{12} = \frac{12}{12} - \frac{5}{12} = \frac{7}{12}$ were girls

5. Pizza Ethan got: $\frac{5}{8} - \frac{3}{8} = \frac{2}{8}$ Ethan and Jaden got: $\frac{2}{8} + \frac{2}{8} = \frac{4}{8}$ Fraction on the table: $1 - \frac{4}{8} = \frac{8}{8} - \frac{4}{8} = \frac{4}{8}$ or $\frac{1}{2}$

6. All the girls got: $\frac{3}{9} + \frac{5}{9} = \frac{8}{9}$ Left for mama: $1 - \frac{8}{9} = \frac{9}{9} - \frac{8}{9} = \frac{1}{9}$

7. The cost has to be shared equally among 4 persons, so we divide the cost by 4 using the fraction $\frac{1}{4}$.

$\frac{1}{4} + \frac{1}{4} + \frac{1}{4} + \frac{1}{4} = 1$

13.8 and 13.9 Practice questions

1) $\frac{5}{3}$ 2) $\frac{14}{5}$ 3) $\frac{73}{9}$ 4) $\frac{47}{8}$ 5) $\frac{24}{5}$ 6) $\frac{64}{10}$ 7) $\frac{68}{9}$ 8) $\frac{48}{5}$ 9) $\frac{39}{8}$ 10) $\frac{79}{12}$ 11) $\frac{37}{7}$

12) $\frac{19}{11}$

13.10 Practice questions

1) $1\frac{1}{2}$ 2) $3\frac{1}{2}$ 3) $2\frac{1}{4}$ 4) $2\frac{3}{5}$ 5) $1\frac{1}{4}$ 6) $3\frac{3}{7}$ 7) $1\frac{8}{9}$ 8) $5\frac{3}{5}$ 9)

13 $\frac{1}{3}$

13.11 Practice questions

1) $4\frac{1}{3}$ 2) 6   3) $7\frac{1}{2}$ 4) $12\frac{2}{3}$ 5) 10   6) $8\frac{2}{9}$ 7) $13\frac{1}{5}$ 8) $12\frac{1}{2}$ 9)

16 $\frac{1}{2}$

13.12 Practice questions

1) $1\frac{1}{2}$ 2) $3\frac{1}{2}$ 3) $2\frac{1}{4}$ 4) $2\frac{3}{5}$ 5) $1\frac{1}{4}$ 6) $3\frac{3}{7}$ 7) $1\frac{8}{9}$ 8) $5\frac{3}{5}$ 9)

13 $\frac{1}{3}$

Chapter 14

Prior learning

1. Perimeter = 26 cm

14.1 & 14.2 Practice questions

1.

List	Unit of length	Unit of area
a. mm, mm^2	mm	mm^2
b. cm^2, km	km	cm^2
c. m, km^2	m	km^2
d. m^2, cm	cm	m^2
e. mm^2, dm, m^2, hm, km^2	dm, hm	km^2, mm^2, m^2
f. cm^2, m, dm^2, m^2	m	dm^2, cm^2, m^2
g. dam, hm^2, mm, cm^2, m^2, cm, mm^2, m	dam, mm, cm, m	hm^2, cm^2, m^2, mm^2

2. b. m^2

14.3 Practice questions

1. a) 18 cm   b) 16 cm   c) 25 cm   d) 36 m   e) 24.5 cm

2. a) 16 cm   b) 26 cm   c) 26 cm   d) 36 cm   e) 70 cm   f) 76 cm   g) 12 cm

h) 40 cm   i) 75 cm

14.4 to 14.6 Practice question

1.

Count = 21 cm^2 Area of grid = row x column = 7 cm x 3 cm = 21 cm^2	Count = 24 cm^2 Area of grid = row x column = 4 cm x 6 cm = 24 cm^2	Count = 20 cm^2 Area of grid = row x column = 4 cm x 5 cm = 20 cm^2

2. a. Perimeter of the garden = total distance around the garden

$$= 4 m + 8 m + 9 m + 4 m + 8 m + 9 m$$
$$= 42 m$$

  b. m^2

  c. Area of pepper = 3 m x 8 m = 24 m^2 d. Area of corn section = 6 m x 8 m = 48 m^2

  e. Area of onions = 3 m x 4 m = 12 m^2

    Area of the tomatoes = 4 m x 6 m = 24 m^2

    So, the tomatoes cover a larger area. The area of the tomato is $\frac{24}{2}$ = 2 times the onion.

3. Area = 4 full squares + 4 partially covered squares = $4 + \frac{4}{2} = 4 + 2 = 6$. Area is approximately 6 cm^2

4. Area = 12 full squares + 16 partially covered squares = $12 + \frac{16}{2} = 12 + 8 = 20$. Area is approximately 20 cm^2

5. Area = 8 full squares + 30 partially covered squares = $8 + \frac{30}{2} = 8 + 15 = 23$. Area is approximately 23 cm^2

6. & 7. Answers will vary.

8. Perimeter = 2 + 2 + 10 + 10 = 24 cm. Area = 2 x 10 = 20 cm^2

9. Area = 16 full squares + 16 partially covered squares = $16 + \frac{16}{2} = 16 + 8 = 24$. Area is approximately 24 cm^2.

10–13. Answers will vary: Teacher supervised activity.

14. Surface area (S.A.) of cuboid = S.A. of short side x 2 + S.A. of long side x 2 + S.A of top and bottom

$$= (6 \times 4) \times 2 + (8 \times 4) \times 2 + (6 \times 8)\,2$$
$$= 48 + 64 + 96$$
$$= 208 \text{ cm}^2$$

15. Surface area (S.A.) of cuboid = S.A. of long side x 2 + S.A. of short side x 2 + S.A of top and bottom

$$= (7 \times 15) \times 2 + (9 \times 15) \times 2 + (9 \times 7)\,2$$
$$= 210 + 270 + 126$$
$$= 606 \text{ cm}^2$$

Chapter 15

Prior learning

1.

2. a. Circle   b. Square   c. Rectangle   d. Triangle   e. Hexagon   f. Pentagon
   g. Diamond h. Oval

3. a. triangular-based prism   b. square-based prism   c. cube

4. a. AB = Diameter   b. OB = Radius   c. OA = OB = Radii of the same circle

15.1 Practice questions

Shapes to be coloured: Numbers 1, 2, 4, 5, 8, 9, 12, 14 and 15.

15.2 Practice questions

1.	2.	3.	4.	5.	6.

15.3 and 15.4 Practice questions

1.

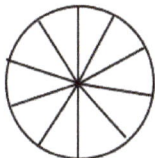

All the lines are the same length.

15.5 Practice questions

1.

a.	1	b.	1	c.	0	d.	2	e.	1
f.	0	g.	0	h.	0	i.	1		

2.

A 1	B 1	C 1	D 1	E 1	F 0	G 0	H 2	I 2	J 0
K 1	L 0	M 1	N 0	O 2	P 0	Q 0	R 0	S 0	T 1
U 1	V 1	W 1	X 2	Y 1	Z 0				

CHEETAH™
Connect to **H**igher **E**ducation, **E**lectronic **T**ools, **A**plication and **H**elp

Chapter 16

Prior learning

1. 10 columns, 3 rows

2. Shortest: 7 boxes

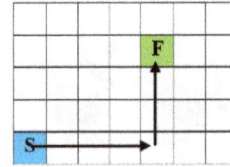

16.1 Practice questions

1.

Location on number grid					
Columns	Rows	Answer	Columns	Rows	Answer
A	1	91	C	9	13
D	6	44	F	2	86
C	4	63	J	2	80
H	5	58	E	1	95
J	7	50	B	5	52
F	9	16	E	6	42
G	2	87	A	3	71
H	10	8	G	7	37
I	3	79	D	10	4
J	8	30	B	4	62

Locations of numbers on grid							
Number	Location	Number	Location	Number	Location	Number	Location
3	C 10	22	B 8	12	B 9	82	B 2
84	D 2	16	F 9	33	C 7	50	J 6
53	C 5	31	A 7	94	D 1		
79	I 3	100	J 1	1	A 10		
46	F 6	36	F 7	28	H 8		
58	H 5	17	G 9	77	G 3		

16.2 Practice question

1a) slide   b) flip   c) slide   d) slide   e) slide   f) flip   g) slide   h) slide   i) turn and slide

2. a) Horizontally: 2 right   Vertically: 4 down   b) Horizontally: 5 right
Vertically: 0
   c) Horizontally: 6 right   Vertically: 0 down   d) Horizontally: 5 right
Vertically: 2 up
   e) Horizontally: 5 right   Vertically: 4 up   f) Horizontally: 3 right
Vertically : 1 up

3. a) 4C   b) 3D   c) 2A   d) 3B   e) 4B

16.3 Practice question

flip of rectangle ABCD across the line shown	flip of triangle XYZ across the line shown	turn around point Y of triangle XYZ
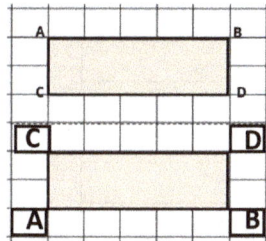	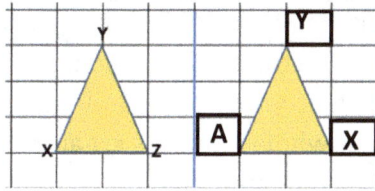	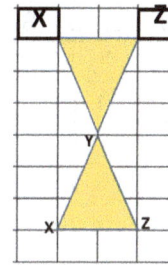
rotations about the blue dot	**turn around point P of triangle PQR**	**slide to the right then up**
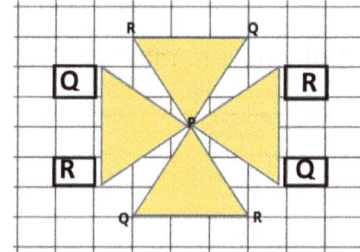	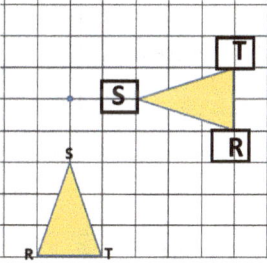	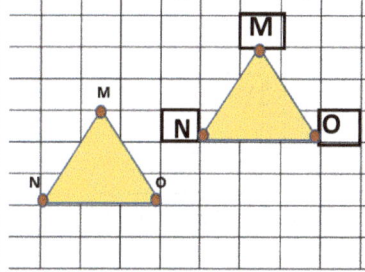

Chapter 17

Prior learning

a. Each number is one more than the number before.

b. Each number is one less than the number before.

17.1 Practice questions

1.

	Pattern	Rule	Next term
a.	4, 8, 12, 16, 20, ...	Add 4 to each number starting at 4	24
b.	10, 20, 30, 40, 50, ...	Add 10 to each number starting at 10	60
c.	23, 21, 19, 17, 15, ...	Subtract 2 from each number starting at 23	13
d.	7, 14, 21, 28, ...	Add 7 to each number starting at 7	35
e.	18, 27, 36, 45, ...	Add 9 to each number starting at 18	54
f.	55555, 4444, 333, 22, ...	Each number is written to its face value number of times starting at 5.	1
g.	32, 27, 22, 17, ...	Subtract 5 from each number starting at 32	12
h.	7, 11, 13, 17, 19, ...	Prime numbers starting at 7	23
i.	97, 95, 93, 91, ...	Subtract 2 from each number starting at 97	89
j.	41, 45, 44, 48, 47....	Add 4 then take away 1 starts at 41	57

CHEETAH
Connect to Higher Education, Electronic Tools, Aplication and Help

2.

	Pattern	Rule
a.	6, 8, 10, 12	List even numbers starting at 6,
b.	9, 12, 15, 18	List multiples of 3 starting from 9
c.	1, 2, 3, 4, 6, 12	List factor of 12 in ascending order.
d.	1, 5, 9, 13	List alternate odd numbers
e.	5, 6, 7, 8	List whole numbers greater than 4
f.	50, 25, 12 ½ , 6 ¼	List repeated division of 100 by 2
g.	1, 4, 9, 16	List squares of numbers 1 to 4
h.	1, 3, 6, 9	List whole numbers multiplied by 3 starting at 1.
i.	15, 12, 9, 6	List of whole numbers decreasing by 3 from 15
j.	18, 25, 32, 39	List of whole numbers increasing by 7 from 18.

3.

A. Add the next number in the series	B. Add the next two numbers in the series
i. 2, 4, 6, **8.**	i. 5. 10, 15, 20, **25**, **30**.
ii. 1, 3, 5, 7, **9.**	ii. 20, 16, 12, 8, **4**, **0** .
iii. 1, 2, 4, 8, **16**.	iii. 12, 18, 24, 30, **36**, **42**.
iv. 3. 6, 9, 12, **15.**	iv. 40, 35, 31, 28, **26**, **25**.
v. 10, 15, 20, **25.**	v. 32, 38, 43, 47, **50**, **52.**
vi. 1, 5, 9, 13, **17**.	vi. 1, 4, 9, 25, **36**, 49.
vii. 2, 5, 9, 14, **20**.	vii. 1, 2, 5, 10, **17**, **26.**
viii. 20, 16, 13, 11, **10.**	viii. 19, 14, 10, 7, **5**, **4**.
ix. 24, 19, 14, 9, **4** .	ix. 30, 35, 29, 34, **28**, **33**.
x. 20, 24, 23, 27, **26**.	x. 32, 26, 21, 17, **14**, **12.**

4.

i. A, C, E, G, **I**.	v. XW, TS, PO, LK, **HG**.	viii. W, S, O, K, **G**.
ii. Z, X, V, T, **R**.	vi. BA, DC, FE, HG, **JI**.	ix. BA, ED, HG, KJ, **NM**.
iii. ZA, YB, ZC, WD, **VE**.	vii. D, H, L, P, **T**.	x. WA, WD, WG, WJ, **WM**.
iv. CD, GH, KL, OP, **ST**.		

17.2 Practice questions

1.

Number of triangles	1	2	3	4	5	6	7	8	9	10	pattern: multiples of 3
Number of edges	3	6	9	12	15	18	21	24	27	30	

2.

   1   2   3   4

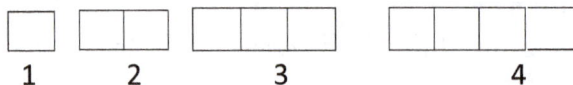

Increasing the number of boxes by one each time.

3.

a.	
b.	
c.	
d.	
e.	

4.

| a. | b. | c. | d. | e. |
| f. | g. | h. | i. | j. |

17.3 Practice questions

1A. i) 2 + 3 + 4 + 5 = 14   ii) ○▲○▲○▲   iii) 6 + 7 + 8 = 21

   iv) 9 shapes   v) 5 circle   vi) 4 triangles

B.

Box number	Number of circles	Number of triangles	Total number of shapes
1	1	1	2
2	2	1	3
3	2	2	4
4	3	2	5
5	3	3	6
6	4	3	7
7	4	4	8

8	5	4	9

2.a. Add one box to the right and one down   b. 1, 3, 5, 7

c.

d. 9, 11, 13, 15 and 17

3.

Ordinal numbers	Geometric shapes	Number of red dots	Number of houses	Number of sticks
1st		5	1	6
2nd		10	2	12
3rd		15	3	18
4th		20	4	24

Chapter 18

Prior learning

1. a. Four take away two   b. Twelve plus three equal fifteen
    c. The product of five and six   d. Twenty divided by 5
    e. Three is added to thirteen and nine taken from the sum
2. a. 14 + 2   b. 12 − 3   c. 2 x 8   d. 8 ÷ 4   e. 4 x n + 9
3. a. 10 + 6   b. 12 − 8   c. 4 − n   d. 4 + n   e. n- 4

18.1 Practice questions

1.

$50	+	$20	+	$20	+	$10	=	$100
$20	+	$50	+	$10	+	$20	=	$100
$10	+	$20	+	$50	+	$20	=	$100

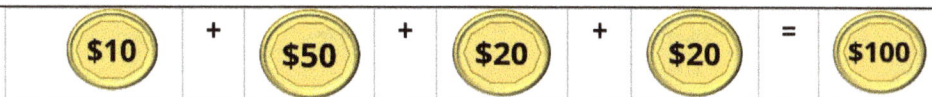

| $10 | + | $50 | + | $20 | + | $20 | = | $100 |

18.2 Practice question

1. B + 5
2. 3 x X = 3X
3. $\frac{1}{2}$ x r or $\frac{r}{2}$
4. B – 6
5. $C + $80
6. $\frac{D}{2}$ + 8
7. 4x – 8
8. y + 6 minutes
9. X – 8 minutes
10. n + n + 2 minutes

18.3.1 Practice questions

1. 4 x 35 = 140 tickets
2. 40 x $9 = $360
3. 100 x 40 x 5 = 500 x 40 = $20,000
4. 35 – 26 = 9 grapes
5. 84 ÷ 6 = 14 students
6. $85 ÷ 5 = 17 students
7. Collected in all = $85 + $5 = $90. So, $90 ÷ $15 = 6 students
8. Ticket sold $65 ÷ $5 = 13 tickets. Not sold = 20 – 13 = 7 tickets
9. Shortfall = $9000 - $6000 = $3000 needed. From each student = $3000 ÷ 30 = $100

18.3.2 Practice questions

1. x + 2 = 8 X + 2 – 2 = 8 – 2 X = 6	2. 2p + 3 = 7 2p + 3 - 3 = 7 - 3 2p = 4 $\frac{2p}{2} = \frac{4}{2}$ p = 2	3. 2n – 8 = 12 2n – 8 + 8 = 12 + 8 2n = 20 $\frac{2n}{2} = \frac{20}{2}$ n = 10	4. $\frac{T}{3}$ + 9
5. 4x + 1 = 9 4x + 1 - 1 = 9 – 1 4x = 8 $\frac{4x}{4} = \frac{8}{4}$ x = 2	6. 3n – 4 = 5 3n – 4 + 4 = 5 + 4 3n = 9 $\frac{3n}{3} = \frac{9}{3}$ n = 3	7. $\frac{4x}{3}$ = 15	8. ½ of x = 8 x = 8 x 2 x = 16
9. 4 x t = 20 $\frac{4t}{4} = \frac{20}{4}$ t = 5	10. 2 x Mika + $10 = Xavier; Meka = $20 2 x $20 + $10 = $50 Xavier spent $50.		

18.3.3 Practice question

A.

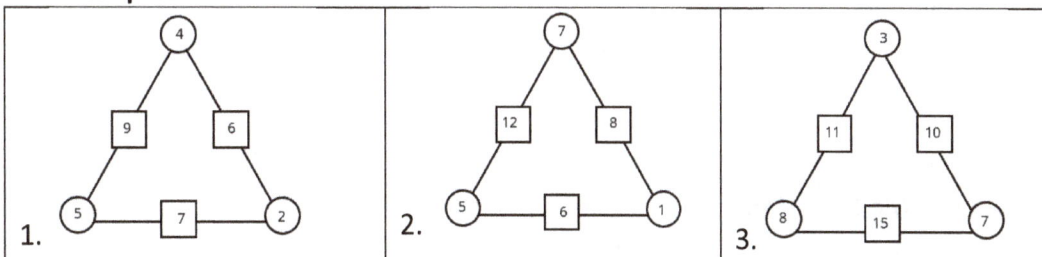

1.
2.
3.

4.

Triangle: 5 (top), 12, 14, 7, 16, 9
5+7=12
7+9=16
9+5=14

5.

Triangle: vertex (top), edge, edge, vertex, edge, vertex

B.

1. Triangle: 5 (top), 7, 11, 2, 8, 6

2. Triangle: 7 (top), 13, 12, 6, 11, 5

3. Triangle: 7 (top), 18, 11, 11, 15, 4

4. Triangle: 13 (top), 25, 28, 12, 27, 15

5. Triangle: 6 (top), 15, 19, 9, 22, 13

6. Triangle: 1 (top), 6, 9, 5, 13, 8

7. Triangle: 6 (top), 15, 20, 9, 23, 14

8. Triangle: 28 (top), 47, 41, 19, 32, 13

9. Triangle: 17 (top), 28, 19, 11, 13, 2

Chapter 19

Prior knowledge:
1. a. day 4   b. 100 pages   c. day 1   2. a. 100   b. 260 oranges   c. oranges

19.1 Practice questions

Mean	Median	Mode		Mean	Median	Mode		Mean	Median	Mode
5	4.5	4	v.	14	32	none	ix.	49	45	45
5	5	5	vi.	24.5	24	24	x.	25	24	24
51	50.5	none	vii.	20	23	24				

| 15 | 14 | 13 | viii. | 71 | 73 | 73 | | | |

19.2 Practice questions

Question number	Mean average	Number of addends	Total	Question number	Mean average	Number of addends	Total
i.	18	3	54	vi.	12	2	24
ii.	8	4	32	vii.	16	5	80
iii.	8	7	56	viii.	32	4	128
iv.	12	5	60	ix.	15	5	75
v.	15	8	120	x.	25	4	100

19.3.1 Practice question

1.

a. Sum of 4 numbers 5 x 4 = 20 Sum of numbers given 2 + 4 + 8 = 14 Missing = 20 − 14 = 6	b. Sum of 4 numbers = 4 x 9 = 36 Sum of given 8 + 12 + 10 = 30 Missing number = 36 − 30 = 6
c. Sum of 5 numbers = 5 x 8 = 40 Sum of given = 12 + 9 + 5 + 13 = 39 Missing number = 40 − 39 = 1	d. Sum of 4 numbers = 4 x 8 = 32 Sum of 3 given = 9 + 11 + 7 = 27 Missing = 32- 27 = 5
e. Sum of 4 numbers = 4 x 25 = 100 Sum of 3 given = 14 + 36 + 12 = 62 Missing number = 100 − 62 = 38	f. Sum of 5 number = 5 x 40 = 200 Sum of 4 given = 26 + 68 + 57 + 18 = 169 Missing number = 200 − 169 = 31
g. Sum of 9 numbers = 9 x 12 = 108 Sum of 9 given = 4 + 9 + 11 + 17 + 21 + 14 + 3 + 13 = 92 Missing number = 108 − 92 = 16	h. Sum of 3 numbers = 3 x 38 = 114 Sum of 2 given = 58 + 47 = 105 Missing number = 114 − 105 = 9
i. Sum of all work = 4 x 10 = 40 hrs Sum of numbers given = 8 + 12 + 10 = 30 Thursday missing day = 40 − 30 = 10	j. Sum of cost for 5 sweets = 5 x 20 = $100 Sum of sweets given = 30 + 25 + 10 + 15 = $80 Missing cost $ 100 - $80 = $20

2.

a. 4 sweets	b. Wednesday	c. 30 sweets ÷ 6 days = 5 sweets per day
d. 9/2 = 4 1/2 = 5 sweets	e. 5 sweets	

3.

a. week 4	b. total = 9 + 5 + 3 + 12 + 4 + 8 = 41 boxes	c. $\frac{41}{6}$ = 6.8	d. 5 boxes

4.

a. Monday	b. 3 + 5 + 10 + 12 + 4 + 8 = 42 hours	c. Means is $\frac{42}{6}$ = 7 hours	d. Median is $\frac{13}{2}$ = $6\frac{1}{2}$ hours

5.

a. Action movies	b. Horror movies	c. Mean = $\frac{3+5+2+4+11+5}{6} = \frac{30}{6} = 5$ teachers	d. $\frac{4+5}{2} = 4\frac{1}{2}$ = 5 teachers	e. Action movie

6.

a. vanilla	b. no	c. Total = (3 + 5 + 10 + 12 + 11 + 8) x 2 = 98	d. chocolate cookie	e. This survey could be used to select the most likely thing to sell for a fundraiser.

7.

a. beach & music	b. no	c. $\frac{6+20+18+4+20+16}{6} = \frac{84}{6} = 14$	d. $\frac{8+9}{2} = 8\frac{1}{2}$	e. Beach, music, hiking, rafting

8.

a. Stew chicken	b. 20	c. $\frac{9+15+27+6+12+21}{6} = \frac{90}{6} = 15$	d. $\frac{70}{2} = 35\frac{1}{2}$	e. Stew chicken

9.

a. Textbooks	b. no	c. Mean = $\frac{3+5+9+2+4+7}{6} = \frac{30}{6}$	d. Median = $\frac{12+15}{2}$ = 13.5	e. Textbook, novels, old newspaper

10.

a. Mangoes	b. Mean = $\frac{3+5+2+10+9+7}{6} = \frac{36}{6} = 6$	c. Median = $\frac{5+7}{2} = 6$	d. apples, oranges and watermelons

11.

a. swimming	b. no	c. Mean = $\frac{30+50+20+40+110+50}{6} = \frac{300}{6} = 50$	d. Median = $\frac{40+50}{2} = 45$	e. Tracks with 120 students

12.

a. Mean = $\frac{8+7+8+9+7+6+4+7}{8} = \frac{56}{8} = 7$ marks	b. yes	c. 7 marks

13.

a. 5	b. No. The most likely number is the mode average which is 5.	c. Answer option C

Chapter 20

Prior learning

1.

Transport	Tally	Frequency
Walk	JHT JHT JHT	15
Bus	JHT JHT II	12
Car	JHT II	7
Bike	JHT I	6

2. a. 2   b. 23   c. 20   d. 10   e. 16

Pencil	Paper	Tally sheet	Questionnaire	Pen	Graph paper	Charts	Survey	Tables
R	C	C	C	R	C	C	C	C

3.

4.

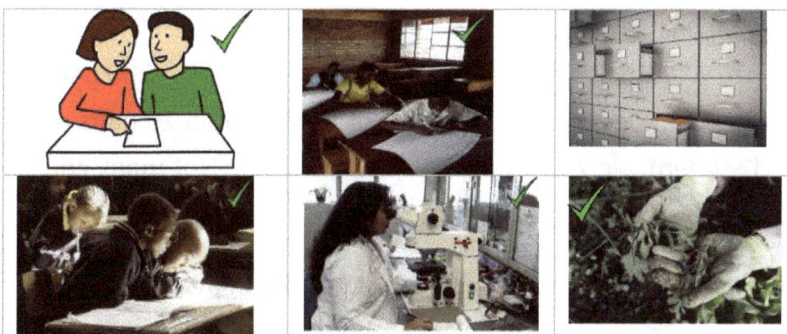

5. X – sample is taken from the population   6. (d) same number of each type of individual.

20.1 Practice questions

1. a. 14 + 10 + 4 + 6 + 2 = 36 pets   b. dog   c. 7 times   d. $\frac{10}{40} = \frac{1}{4}$ e. 40 – 36 = 4 students had no pet

2. a. shop 3   b. shop 5   c. 40 - 10 = 30 computers.   d. 15 + 25 + 40 + 25 + 10 = 115 computers
   e. shop 2 and 4

3. a. 3 + 5 + 8 + 5 + 2 + 7 + 4 + 6 + 3 + 9 + 1 + 3 = 56 students   b. October   c. November
   d. 10 students   e. $\frac{8}{56} = \frac{1}{7}$

4. a. 3 + 5 + 8 + 5 + 2 + 7 = 30 students   b. gym   c. 8 – 2 = 6 students   d. Track and field , karate
   e. music and drama

5a. 60 + 50 + 80 + 50 + 20 + 70 = 330 animals   b. Ducks   c. pig   d. 80 – 20 = 60   e. 165 animals

20.1.2 Practice question

1. a. 90   b. Student 5 (S5)   c. 20   d. S3 and S8 and they got a score of 60

CHEETAH
Connect to Higher Education, Electronic Tools, Aplication and Help

e. Mean average is $\dfrac{20+50+60+40+90+70+80+60+70+50}{10}$ = 59

2. a. 18 books b. Thursday c. Wednesday and Saturday   d. 18 - 8 = 10 books

   e. Mean = $\dfrac{18+14}{2}$ = 16 books

3. a. 9 goals in match #5   b. 4 goals in match #4   c. 8 + 6 + 7 + 4 + 9 + 7 = 41 goals

   d. matches 3 and 6   e. $\dfrac{21}{3}$ = 7 goals

4. a. 8 cars   b. week 3   c. 6 + 8 + 3 + 6 + 4 + 7 = 34 cars   d. going up and down   e. weeks 4 and 6

5. a. 8 leaves   b. between days 3 and 4   c. between days 6 and 8   d. 3 leaves fell off e. 5 leaves

20.1.3 Practice question

1. a. 14 grasshoppers   b. 20 − 4 = 16   c. 12 − 4 = 8   d. Total is 14 + 4 + 10 + 20 + 12= 60

   e. The neighbour's yard; Birds would be most likely attracted to the many grasshoppers in the

   neighbour's yard.

2. a. Devin took 5 apples   b. Dave took the most apples   c. Desmond took the least 4 apples

   d. 7 + 5 + 4 + 8 + 9 = 33 apples   e. 8 apples

3 a. 70 pairs of shoes   b. Walk in Style   c. 30 + 40 + 70 + 40 + 20 = 200 pairs of shoes

   d. Average = $\dfrac{200}{5}$ = 40 pairs of shoes   e. Discount Centre; Sold 70 pairs of shoes

4 a. Little Mermaid   b. 1 student   c.7 students   d. Little Critters. Not many children would attend

   e. 40 book reports

5 a. Parking lot B   b. 60 cars   c. 8 cars   d. parking lot D e. Mean average is $\dfrac{6+24+12+10+8}{5}$ = 12 cars

20.1.4 Practice questions

1 a. cows   b. $\dfrac{1}{8}$ c. almond milk and soy milk d. $\dfrac{40}{4}$ = 10 students   e. 0 students

2a. 20 students   b. Ms Gloke and Ms Gnome   c. $\dfrac{1}{4}$ d. 20 − 5 = 15   e. Ms Snook as she got 1/2 all the votes.

3 a. horror   b. 25 students   c. $\dfrac{1}{8}$ d. Action and comedy   e. Comedy or action

4 a. ½   b. 50 children   c. 50 children   d. 20 - 10 = 10 children

e.

Traditional snack	Fraction	Number of children	Traditional snack	Fraction	Number of children
Coconut drops	$\dfrac{2}{10}$	20	Peanut cake	$\dfrac{5}{10}$ or ½	50
Gizzard	$\dfrac{1}{10}$	10	Jackass corn	$\dfrac{1}{10}$	10
Grater cake	$\dfrac{1}{10}$	10			

5. a. Rent and bills   b. $20   c. $\frac{2}{10}$

d.

Expenses	Fraction	Number of children	Expenses	Fraction	Number of children
Rent and bills	$\frac{5}{10}$ or $\frac{1}{2}$	$50	Children	$\frac{1}{10}$	$10
Food	$\frac{2}{10}$	$20	Saving	$\frac{1}{10}$	$10
Medical expenses	$\frac{1}{10}$	$10			

e. No. All the money was spent.

20.2 Practice questions

1. Bar chart showing favourite types of milk chosen by a class of 40 students

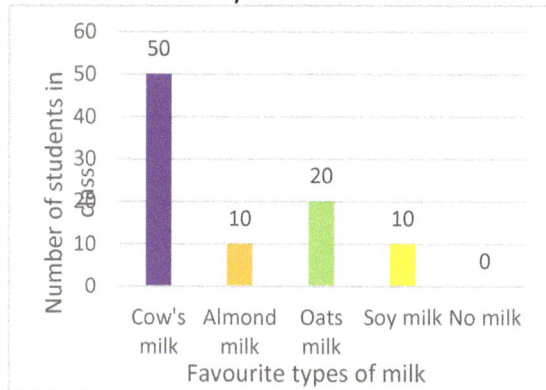

2. Bar chart of favourite drama club teachers

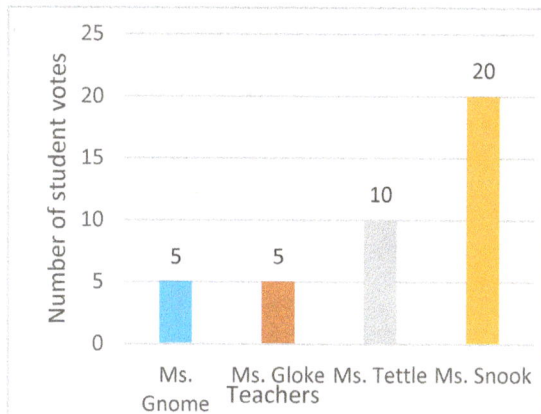

3. Pie chart of types of movies students disliked

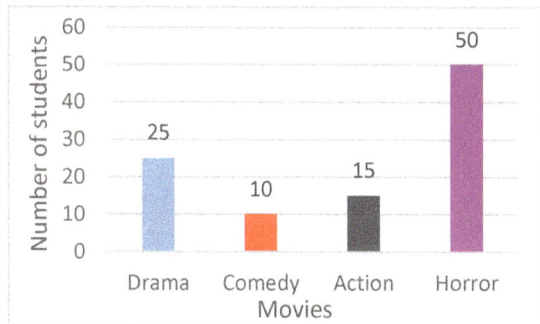

4. Bar chart of children who could correctly identify traditional Jamaican snacks from a group of 100

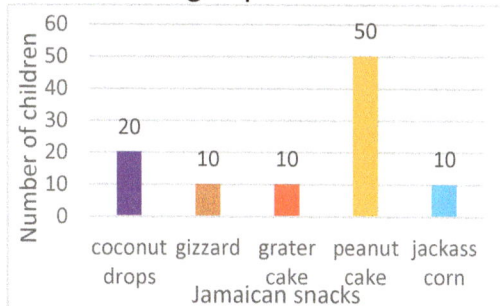

5. Bar chart showing how Janeil's mother spent her salary of $100 for the week

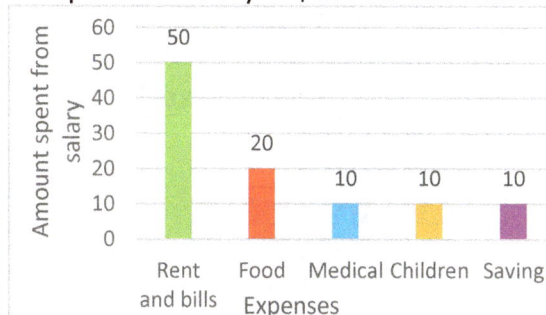

20.3.1 Practice questions

1. Bar chart showing the number of grasshoppers seen at different places

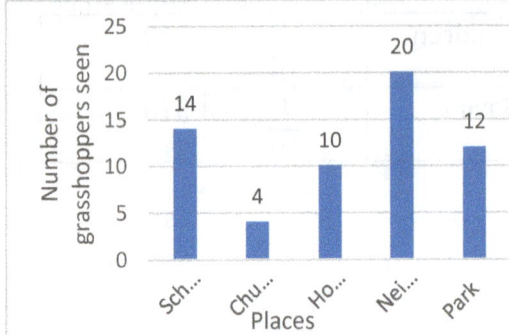

2. Bar chart shows the number of apples received by each boy.

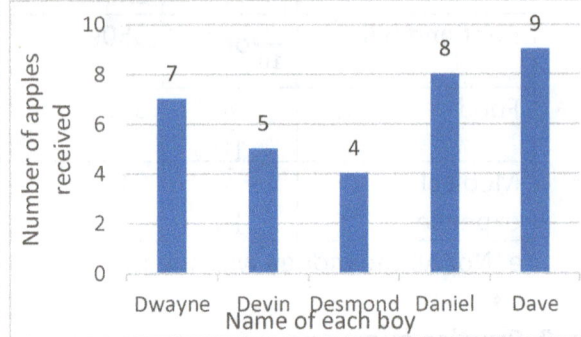

3. Bar chart showing the number of pairs of shoes sold at each shoe store.

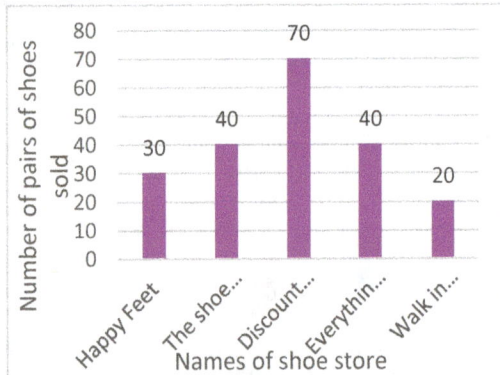

4. Bar chart showing the number of book reports written about these movies.

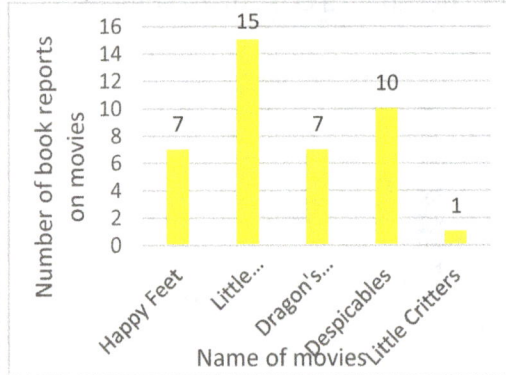

5. Bar chart showing the number of cars parked in five parking lots.

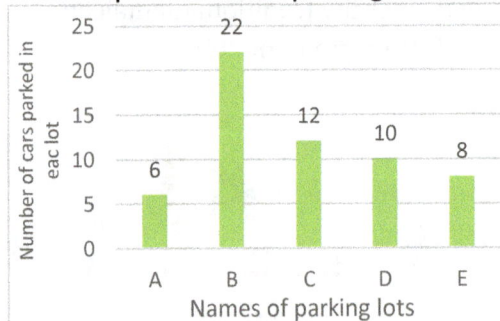

20.3.2 Practice questions

1. Pictograph showing the number and type of pet children in a grade four class have at home.

Animals	Pets
Dogs	
Cats	

2. Pictograph showing the number of computers sold at each of five shops at a technology fair.

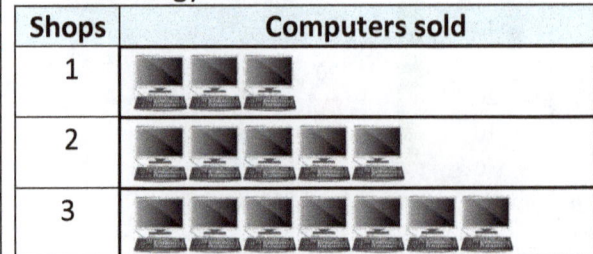

Shops	Computers sold
1	
2	
3	

Fish	
Bird	
Lizard	

Key: Each colour block = __2__ animals

4	
5	

Key: Each computer represents
__5__ computers

3. Pictograph showing the number of students in a class born in each month.

Month	Students
January	
February	
March	
April	
May	
June	
July	
August	
September	
October	
November	
December	

Key: Each person represents **1 student born in each month.**

4. Pictograph of student participation in clubs in a grade 4 class.

Clubs	Student participation
Football	
Track and field	
Gym	
Karate	
Drama	
Music	

Key: Each person represents **1 student in each club.**

5. Pictograph showing the number of animals on a farm.

Animals	Number on a farm	Animals	Number on a farm
Cow		Fowl	
Goat		Pig	
Duck		Sheep	

Key: Each animal = **10 animals**

Chapter 21

Prior learning
1. a. Add(+) b. Subtract (-) c. multiply (x) d. divide (÷) e. Add (+) then subtract (-)
   f. Multiply (x), add (+) then subtract (-)
2. a. 6 x 4 b. 1, 2, 3, 4, 6, 8 c. 7 x 3 d. 8 x 2 and factors 3, 5, 6 and 7
e. 6 x 7 = 42 blocks in the array; 3, 4, 5 and 8 could not be a factor of 42.
3. 20 ÷ 3 = 6 persons can get 3 chocolate blocks. Two (2) chocolate blocks are left over.

1. 6 x 8 = 48 litres   2. 28 x 7 = 196 cars   3. 54 ÷ 6 = 9 days   4. 24 − 9 = 15 litres
5. Leon has 24 phone calls, Beverly has 24 + 14 = 38 calls; Altogether 24 + 38 = 62 calls.
6. 42 − 27 = 15 people   7. 57 − 19 = 38 early students
8. 104 + 89 = 193 people; 268 − 193 = 75 children
9. Mary took 38 + 27 = 65 minutes   10. 234 − 185 = 49 has no seat

21.1 Practice questions
1. 78 ÷ 6 = 13 = quotient   2. remainder = 3   3. 9 x 8 = 72   4. 14
5. 8 x 24 + 3 = 192 + 3 = 195   6. 37 x 8 = 296 used up; 300 − 296 = 4 nails left over
7. 371 ÷ 8 = Each participant paid $46 with $3 change
8. 9 x $90 = $810; They need $500 more.
9. 260 ÷ 8 = 32 and remainder 4 pages   10. 189 ÷ 5 = 37 remainder 4

21.2 Practice questions
1.
1. 219	2. 141	3. 89 R3	4. 445 R3
5. 752	6. 81 R5	7. 132	8. 659 R 6
9. 490 R2	10. 40	11. 259 R11	12. 141

21.3 Practice questions
1.
a. 307	b. 206	c. 1045	d. 2065
e. 506	f. 601	g. 809	h. 9076

2.
a. 150	b. 820	c. 50	d. 210	e. 700
f. 403	g. 102	h. 603	i. 705	j. 406

21.4 Practice questions

Item	Divisible by	Numbers
a.	2	24, 66, 128, 150, 184, 472, 592, 502, 594, 406, 472, 506, 358
b.	3	24, 48, 66. 78, 150,
c.	4	6, 24, 184, 472

2.

Number	2	3	4	none
24	✓	✓	✓	
96	✓	✓	✓	
48	✓	✓	✓	
72	✓	✓	✓	

CHEETAH
Connect to Higher Education, Electronic Tools, Aplication and Help

763				✓	
1104	✓	✓	✓		
153		✓			
203				✓	
748	✓				
852	✓				

21.5 Practice questions

a. $1409\frac{1}{7}$	b. $934\frac{2}{5}$	c. $459\frac{3}{8}$	d. $1030\frac{5}{9}$	e. $1418\frac{1}{4}$	f. $375\frac{2}{3}$
g. $749\frac{10}{11}$	h. $274\frac{1}{2}$	i. $721\frac{1}{3}$			

21.6 Practice questions

1. 12 in the wrong position
2. 4 x 9 is not 32
3. 3 x 7 is not 14
4. no zero needed at front; 4 x 8 is not 28
5. 3 into 7 is not 3
6. 7 into 48 is not 9

Chapter 22

Prior learning

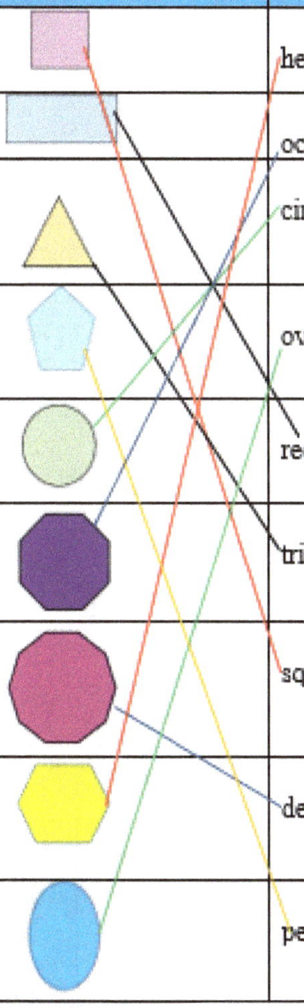

Plane shape	Name
	hexagon
	octagon
	circle
	oval
	rectangle
	triangle
	square
	decagon
	pentagon

CHEETAH™
Connect to Higher Education, Electronic Tools, Aplication and Help

22.1 Practice questions

1.

Shape	No. of edges	No. of vertices	No. of faces	Shape of faces
cube	12	8	6	square
cuboid	12	8	6	rectangle

2.

Shape	No. of edges	No. of vertices	No. of faces	Shape of faces
cube	12	8	~~8~~ 6	~~rectangle~~ square
cuboid	~~9~~ 12	~~6~~ 8	~~7~~ 6	~~square~~ rectangle

3. D

22.2.1 Practice questions
1. A   2. A.   3. B.   4. B.

22.2.2 Practice questions
i. YES   ii. YES   iii. YES   iv. YES   v. NO   vi. YES   vii. YES   viii. YES   ix. NO   x. NO   xi. NO   xii. NO

Chapter 23

Prior learning
1. C. 5
2.

Number sentences	Word sentences
$7 + 19 = 26$	If the sum of 2 numbers is eight and one of the numbers is 6, what is the other number?
$2 + 8 = 10$	There are forty students in the class and 28 came early. How many came late?
$25 - 9 =$	Increase 7 by 19 to give 26.
$18 \div 6 = ?$	How many times is 18 more than 6?
$24\,m \div 6 = ?$	There are 8 benches in the class and each bench holds four students. How many students are in the class?
$a + b = 8$ and $a = 6$, Find b	Decrease 25 by 9
$12 \div a = 4$ find a	A piece of thread 24 metres long was cut into 6 equal parts how long was one part?
$8 \times 4 = ?$	Two plus eight is 10.
$40 - 28 = ?$	When 12 apples were shared equally each child got four, how many children got apples?

23.1 Practice questions

1. $5 + 2 - 3$ $= 7 - 3$ $= 4$	2. $7 - 5 + 6$ $= 2 + 6$ $= 8$	3. $4 + 3 \times 2$ $= 4 + 6$ $= 10$	4. $2 \times 3 - 2$ $= 6 - 2$ $= 4$	5. $9 - 2 \times 4$ $= 9 - 8$ $= 1$
6. $5\,(3) - 4$ $= 5 \times 3 - 4$ $= 15 - 4$ $= 11$	7. $6 \div 2 + 7$ $= 3 + 7$ $= 10$	8. $9 - 8 \div 4$ $= 9 - 2$ $= 7$	9. $2 + 3 \times 4 - 1$ $= 2 + 12 - 1$	10. $9 - 4 \div 2 + 1$ $= 9 - 2 + 1$ $= 7 + 1$

CHEETAH
Connect to Higher Education, Electronic Tools, Aplication and Help

			= 14 – 1	= 8
			= 13	
11. 3(5 + 2)	12. 4(6 - 1)	13. (7 - 4) + 2	14. (4 + 2) – 2	15. 4 + (2 – 2)
= 3(7)	= 4(5)	= 3 + 2	= 6 – 2	= 4 + 0
= 3 x 7	= 4 x 5	= 5	= 4	= 4
= 21	= 20			

23.2 Practice questions

1. Isiah = 3 x Craig's age, where Craig = 4 years old; Isiah = 3 x 4 years = 12 years old.	2. Timothy = 5 years + 3 years = 8 years
3. Elaine had 135 sheets + 55 sheets = 190 sheets	4. Alexia got 25 x 2 = 50 sweets. So, Bianca had 86 + 25 + 50 = 161 sweets

5. x is 24 ÷ 8 = 3	6. $p + 8 = 20$ $p + 8 - 8 = 12 - 8$ $p = 4$	7. $5 \times n = 80$ $\frac{5}{5} \times n = \frac{80}{5}$ $1 \times n = 16$ $n = 16$	8. ½ of n = 12 $n = 12 \times 2$ $n = 24$
9. $24 \div n = 6$ $n = \frac{24}{6}$ $n = 4$	10. 6 x 6 = 36 Hence n = 6	11. (n x 2) 2 = 20 (2n) 2 = 20 2n x 2 = 20 4n = 20 $n = 20 \div 4 = 5$	Or double a number, then find three times the answer give 30. Work backwards. 1st take 30 and divide it into 3 = 10. 2nd take 10 and split it by 2 = 5

23.3.1 Practice questions

a. 2b = 2 x 3 b = 6	b. cab = 4 x 2 x 3 cab = 24	c. 3a + 2c + 2b – 2d = 3 x 2 + 2 x 4 + 2 x 3 – 2 x 5 = 6 + 8 + 6 – 10 = 20 – 10 = 10
d. 4a + 4b – 4c = 4 x 2 + 4 x 3 – 4 x 4 = 8 + 12 – 16 = 20 – 16 = 4	e. 2d + 2c + 2b + 2a = 2 x 5 + 2 x 4 + 2 x 3 + 2 x 2 = 10 + 8 + 6 + 4 = 28	f. 7a + 6b + 5c + 4d = 7 x 2 + 6 x 3 + 5 x 4 + 4 x 5 = 14 + 18 + 20 + 20 = 72

23.3.2 Practice questions

1. x + x = 6 2 x = 6 $\frac{2}{2}$ x = $\frac{6}{2}$ x = 3	x + y = 9 3 + y = 9 3 - 3 + y = 9 - 3 y = 6	y + z = 14 6 + z = 14 6 − 6 + z = 14 − 6 z = 8	x + y + z = 3 + 6 + 8 = 17

2. A + A = 4	A + B = 9	B + C = 14	A + B + C
2 A = 4	2 + B = 9	7 + C = 14	= 2 + 7 + 7
$\frac{2}{2}$ A = $\frac{4}{2}$	2 - 2 + B = 9 - 2	7 − 7 + C = 14 − 7	= 16
A = 2	B = 7	C = 7	

3.

+ = $50,

= $50 ÷ 2

= $25

+ = $ 75

$25 + = $75

= $75 - $25

= $50

+ = $80

$50 + = $80

= $80 - $50

= $30

+ +

= $25 + $50 + $30

= $105

4. Since the hanger is not mending to one side then each side of the hanger has the same weight. So 36 g ÷ 2 = 18g.

The left side has	For the right side
Triange + Cube + Drum = 18g	Cylinder + Drum + cube = 18g
So Triangle + 3g + 8g = 18 g	So Cy;inder + 7g + 3g = 18 g
Triangle + 11g = 18g	Cylinder + 10 g = 18 g
Triangle = 18g – 11g	Cylinder = 18g – 10g
Triangle = 7 g	Cylinder = 8g

5.

+ = $10

2 = $10

= $10 ÷ 2

= $5

+ = $15

$5 + = $15

= $15 - $5

= $10

+ = $17

$10 + = $17

= $17 - $10

= $7

+ −

= $5 + $10 + $7

= $ 22

6.

+ = $8

2 = $8

= $8 ÷ 2

= $4

+ = $9

$4 + = $9

= $9 - $4

= $5

+ = $11

$5 + = $11

= $11 - $5

= $6

x +

= $4 + $5 + $6

= $15

7. 🥬 + 🥬 = $20
    2 🥬 = $20
    🥬 = $20 ÷ 2
    🥬 = $10

🥬 + 🍅 = $19
$10 + 🍅 = $19
🍅 = $19 - $10
🍅 = $9

🍅 + 🥕 = $14
$9 + 🥕 = $14
🥕 = $14 - $9
🥕 = $5

🍅 + 🥬 − 🥕
= $9 + $10 - $5
= $19 - $5
= $14

8. 🔋 + 🔋 = $40
    2 🔋 = $40
    🔋 = $40 ÷ 2
    🔋 = $20

🔋 + ✏️ = $70
$20 + ✏️ = $70
✏️ = $70 - $20
✏️ = $50

✏️ + 📏 = $90
$50 + 📏 = $90
📏 = $90 - $50
📏 = $40

✏️ + 🔋 + 📏
= $20 + $50 + $40
= $110

Chapter 24

Prior learning

1. c. chance and e. likelihood

24.1 and 10.2 Practice questions

1. i. $\frac{5}{10}$ or ½   ii. $\frac{4}{10}$ or $\frac{2}{5}$ iii. yellow

2. i. $\frac{6}{26}$ or $\frac{3}{13}$ ii. $\frac{2}{26}$ or $\frac{1}{13}$ iii. $\frac{26}{26}$ or 1 (All the balls are round.) iv. 0 (none of the balls have stripe) v. yellow (2 yellow balls out of 26 = $\frac{1}{13}$)

3. i. six   ii. $\frac{1}{6}$ iii. $\frac{1}{6}$ iv. $\frac{2}{6}$ or $\frac{1}{3}$ (There are 2 numbers greater than 4; 2 out of 6) v. $\frac{1}{6}$

4. i. Total = 50 + 20 + 10 + 10 + 10 = 100 patties   ii. $\frac{20}{100}$ or $\frac{1}{5}$ iii. $\frac{10}{100}$ or $\frac{1}{10}$ iv. 0 (no chance of getting a cheese patty because there are no cheese patties v. Beef patty. They have a $\frac{50}{100}$ or $\frac{1}{2}$ chance of getting a beef patty.   vi. soy, vegetable and shrimp

5. i. $\frac{3}{6}$ or $\frac{1}{2}$ ii. $\frac{2}{6}$ or $\frac{1}{3}$ iii. $\frac{3}{6}$ or $\frac{1}{2}$ iv. $\frac{6}{6}$ or 1

6. i. ½   ii. $\frac{1}{10}$ iii. $\frac{8}{10}$ or $\frac{4}{5}$ iv. $\frac{3}{10}$

7. i. b. Very likely   ii. b. Red   iii. Purple

8. i. $\frac{8}{40}$ or $\frac{1}{5}$ ii. Least likely   iii. d. Red   v. Yellow

9. $\frac{13}{52}$ or $\frac{1}{4}$

10. i. $\frac{3}{8}$ ii. $\frac{2}{8}$ iii. watermelon and fries   iv. $\frac{1}{8}$

CHEETAH™
Connect to Higher Education, Electronic Tools, Aplication and Help